Neutron Scattering Data Analysis 1990

Neutron Scattering Data Analysis 1990

Invited and contributed papers from the conference on Neutron Scattering Data Analysis held at The Rutherford Appleton Laboratory, Chilton, 14–16 March 1990

Edited by M W Johnson

Institute of Physics Conference Series Number 107
Institute of Physics, Bristol, Philadelphia and New York

sep/ue
phys

CODEN IPHSAC 107 1–282 (1990)

British Library Cataloguing in Publication Data

Conference on Neutron Scattering Data Analysis (1990:
  Rutherford Appleton Laboratory)
  Neutron scattering data analysis 1990.
  1. Neutrons. Scattering. Data. Statistical analysis.
  I. Title II. Johnson, M. W. (Michael William) *1944* –
  539.7213

  ISBN 0-85498-066-0

Library of Congress Cataloging-in-Publication Data are available

Published under The Institute of Physics imprint by IOP Publishing Ltd
Techno House, Redcliffe Way, Bristol BS1 6NX, England
335 East 45th Street, New York, NY 10017-3483, USA
US Editorial Office: 1411 Walnut Street, Philadelphia, PA 19102, USA

Printed in Great Britain by Galliard (Printers) Ltd, Great Yarmouth, Norfolk

Preface

Methodology in data analysis may not change as rapidly as the subjects it serves, but it was felt that after four years enough had changed to warrant a second specific meeting on neutron scattering data analysis. This meeting, like its predecessor, was held on 14–16 March 1990 at the Rutherford Appleton Laboratory and the number and level of the contributions clearly demonstrated the value of such a meeting.

The use of maximum entropy techniques is becoming more widespread, while the practitioners of the technique are turning their attention to the difficult question of reliability estimates. Simulated annealing, and the related technique of reverse Monte Carlo, have also made considerable impact on the structural aspects of neutron scattering data analysis and the meeting was intrigued by the real-time analysis of previously unseen data by Bob McGreevy while the meeting progressed.

If a theme was apparent it was perhaps the obvious one that while neutron scattering and computational techniques are individually extremely powerful, the deliberate combination of the two provides an even more powerful tool for probing the condensed state of matter. The new computational techniques also encourage the experimentalist to use *all* of the available information in arriving at a structural or dynamic model.

The meeting was a pleasure to organise, largely because of the excellent administrative assistance provided by Mrs Marjorie Sherwen and Miss Jane Warren. I would also like to thank all the authors for their prompt delivery of manuscripts, and the sponsors of the meeting, BP Research, whose financial support enabled speakers to be invited from overseas.

M W Johnson

Contents

v Preface

Chapter 1: General

1–21 Fundamentals of MAXENT in data analysis
J Skilling and S Sibisi

23–44 Simulated annealing: an introductory review
J Pannetier

45–55 The Bayesian approach to optimal instrument design
D S Sivia, R N Silver and R Pynn

57–67 An amateur's guide to the pitfalls of maximum entropy
A K Soper

69–82 GENIE Version 3: A tool for neutron scattering data analysis and visualization
C M Moreton-Smith

83–91 The visualization of neutron scattering data using UNIRAS
K M Crennell

Chapter 2: Crystalline diffraction

93–100 Extending the power of powder diffraction for structure determination
W I F David

101–116 Maximum entropy reconstruction of spin density maps in crystals from polarized neutron diffraction data
R J Papoular and B Gillon

117–126 Real-time neutron powder diffraction study of $MoO_3 . 2H_2O$ topotactic dehydration
M Anne, N Boudjada, J Rodriguez and M Figlarz

127–134 MXD: A least-squares program for non-standard crystallographic fitting
P Wolfers

135–144 On the normalization of spallation source powder diffraction data
R O Piltz

145–163 The data analysis of reciprocal space volumes
C C Wilson

Chapter 3: Liquids and amorphous diffraction

165–184   Reverse Monte Carlo (RMC) simulation: modelling structural disorder in crystals, glasses and liquids from diffraction data
R L McGreevy, M A Howe, D A Keen and K N Clausen

185–192   A normalization procedure of liquid diffraction data based on the knowledge of the sample neutron cross section
F Cilloco and R Felici

193–211   ATLAS: A suite of programs for the analysis of time-of-flight neutron diffraction data from liquid and amorphous samples
A C Hannon, W S Howells and A K Soper

Chapter 4: Neutron reflectometry

213–222   Analysis of neutron reflectivity data using constrained model fitting
J Penfold

223–231   Maximum entropy analysis of neutron reflectivity data—some preliminary results
D S Sivia, W A Hamilton and G S Smith

Chapter 5: Small angle scattering

233–244   Can we justify conventional SANS data analysis?
R E Ghosh and A R Rennie

Chapter 6: Quasielastic and inelastic scattering

245–252   CLIMAX: A program for force constant refinement from inelastic neutron spectra
G J Kearley and J Tomkinson

253–265   Set-up and optimization of scans with the rotating analyser crystal spectrometer ROTAX
H Tietze, W Schmidt, R Geick, H Samulowitz and U Steigenberger

267–277   Analysis of quasielastic and inelastic scattering from crystal analyser instruments
A Smith, C J Carlile, M Prager and R M Richardson

279   Author Index

281–282   Subject Index

Fundamentals of MAXENT in data analysis

John Skilling   Sibusiso Sibisi*
Department of Applied Mathematics and Theoretical Physics
University of Cambridge
England CB3 9EW

July 1990

Abstract

Neutron diffraction studies, like many others, pose problems of **inference**. The proper tool for analysing these is probability calculus. We stress this point, and show how to incorporate a preference for "smooth" reconstructions into the probabilistic framework. This is illustrated with synthetic interpolation data. Smoothness is incorporated in a natural way, which is easily compatible with maximum entropy analysis. With a firm foundation in probability theory, and the inclusion of smoothness, we expect maximum entropy will become even more useful in the future.

1   Introduction

Neutron Diffraction studies form one among many problems in scientific inference. In this particular example, the basic problem is to infer a geometrical distribution of scattering centres (or at least its radial correlation function) from angular scattering data. Similar problems occur in other areas, such as the inference of a frequency spectrum from time series data, and the inference of a two-dimensional image from interferometer data.

We hold that all such inference problems must, as a matter of principle, be handled using ordinary probability theory. The logical necessity of this view is given in section 2. However, probabilitycalculus gives us a range of answers, and not just a single "best" correlation function, spectrum or image. Of course there is still a "best" result, namely the most probable. It is often legitimate to identify this with the maximum entropy reconstruction. However, there will be a range of plausible reconstructions around the best, which defines the accuracy to which the reconstruction is known. These error estimates are an important part of one's conclusions.

Moreover, the probability distribution of results rests upon a prior probability distribution, which must be assigned before one can do the analysis. Different analysts can and do have different opinions about what this prior should be, and this seems to inject an unwelcome amount of subjectivity into the arguments. However, the probability equations also let us discover how appropriate any particular prior is for the data in question, through a quantity which we call the "evidence". The evidence allows us to discriminate objectively between different proposed priors. In this way, an appropriate degree of objectivity is returned to the analysis.

For intrinsically positive reconstructions such as a scattering correlation function, it is legitimate to use the entropy of a reconstruction to define its prior probability distribution. The "best" result is then the maximum entropy reconstruction; details of this are to be found elsewhere (*e.g.* Skilling [1]). However, that algebra is somewhat complicated by the nonlinear nature of the entropy

---

*Supported by Mobil North Sea Limited

formula, and it is easier to introduce the theory through the limiting case of a Gaussian prior, as we do in this paper.

There have been many successful applications of maximum entropy, but there have also been failures (less publicised for obvious reasons). Chief among these has been the interpolation problem in which the data are localised measurements $f(x_k) = D_k$ of the required function $f(x)$. The entropy prior is uncorrelated in x, so that the maximum entropy reconstruction completely fails to interpolate between the measurement points. Failure in this case exposes imperfections which can sometimes be seen with other types of data.

In section 3, we show how to cure this by using an intrinsic correlation function whcih explicitly codifies our desire for a "smooth" result. Again, the probability formulae let us discriminate among different proposed correlation functions, and we discuss this in section 4.

We expect that the inclusion of intrinsic correlation functions will significantly enhance the power of the maximum entropy technique.

2   Probability Calculus

Scientists simplify, having learned that complicated problems can be broken down into simpler ones. We will follow this methodology, and start at the beginning with general reasoning about simple situations. In all inference problems, we have to consider different propositions. In image reconstruction, these might relate to particular images: thus

> Proposition A: the image is of an aardvark
> Proposition B: the image is of a beagle
> Proposition C: the image is of a cat, etc.

At the very least, we will want to be able to rank our preferences for these different propositions, for example we might

> Prefer B to A, AND Prefer A to C.

Presumably we want these to imply

> Prefer B to C,

otherwise we will soon start to argue in circles. In other words, we need a transitive ranking of preferences. Consequently, we can assign a real number to each preference, slotting in a new number for each new preference depending on where it falls in the existing transitive sequence. Although fairly arbitrary, these numerical codes

$$\pi(\text{Proposition}) \in \mathcal{R}^1$$

are at least correctly ordered. Let us apply this minimal structure to the simplest problems.

1. Start with a "1-bit" proposition $X \in \{\text{True},\text{False}\}$. For example, X = "Confucius was born in 551 BC". X may or may not be true: we don't really know, and we can only express a preference, based as well as may be on such information as we have to hand. Presumably though, our preference for X being False will be determined by our preference for X being True.

> Preference for $\neg X$ $\Leftarrow$ Preference for X.

In terms of our numerical codes, this means that there is some mapping (i.e. a function) f which takes our code for a proposition into the corresponding code for its negation.

$$\pi(\neg X) = f\left(\pi(X)\right)$$

Already, we can start doing some mathematics, because the identity $\neg\neg X = X$ immediately tells us that

$$f(f(x)) = x,$$

so that f is not wholly arbitrary. However, this does not take us very far.

2. Moving on, we take a 2-bit proposition $(U, V) \in$ {TT,TF,FT,FF}. We can reach our joint preference for the two bits (U, V) being, say, TT (both true) in two individual 1-bit steps.

  (a) Express preference for U (say),

  (b) Only if U is true, obtain and use preference for V, given U.

In other words,

$$\text{Preference for } (U, V) \Leftarrow \text{Preference for } U, \text{ Preference for } (V|U)$$

In terms of the codes, there is some function g which generates this joint preference:

$$\pi(U, V) = g(\pi(U), \pi(V|U))$$

Again, we can do a little mathematics, using the Boolean identity $(U, V) = (V, U)$ to reach

$$g(\pi(U), \pi(V|U)) = g(\pi(V), \pi(U|V)),$$

but the structure remains too impoverished to take us very far.

3. Hence we move on to a 3-bit proposition (R, S, T). This can be factored into simpler conditional propositions in half a dozen different ways, such as:

  (a) Express preference for R (say),

  (b) Only if R is true, obtain preference for S (say), given R,

  (c) Only if R and S are both true, obtain preference for T, given R and S.

All of these half a dozen ways must, of course, be equivalent, because the order of R, S, T is irrelevant to their joint truth. The truly remarkable consequence of this elementary observation (Cox [2]), is that there exists some function F of our original rather arbitrary numerical codes π, taking them into other codes

$$\Pr(X) = F(\pi(X))$$

which are distinctly less arbitrary. In fact the new codes obey

$$\Pr(X) + \Pr(\neg X) = 1$$
$$\Pr(X, Y) = \Pr(X)\Pr(Y|X)$$
$$\Pr(\text{False}) = 0 \text{ and } \Pr(\text{True}) = 1.$$

At this point or before, a purist might remember that any preference whatever must be conditional upon some sort of earlier expectation or belief, traditionally given the symbol I, so that we could more properly write

$$\Pr(X|I) + \Pr(\neg X|I) = 1$$
$$\Pr(X, Y|I) = \Pr(X|I)\Pr(Y|X, I)$$

In the interests of notational clarity though, we will often omit this I, which may in any case be a compound proposition common to several steps in an argument.

Anyway, we recognise the standard rules of probability calculus, being used precisely for their original purpose of quantifying our preferences, and we are entirely correct to identify our new codes Pr as probability values. They happen to obey the old-fashioned frequentist definition of probability when that is germane. Indeed, given an infinite sequence of trials, how else could one expect to define one's belief about an arbitrary individual success but on the basis of the overall success ratio? However, our interpretation of probability is far more general. To take the above example, Confucius presumably had only one birthdate, and not an infinite ensemble of them. We can, nevertheless, discuss it probabilistically with full propriety. To follow Jaynes [3] in using the terminology of philosophy, *probabilities are epistemological*, representing our beliefs, rather than ontological, representing objective external reality.

The point being stressed here is not that we can describe our preferences by probability values. The point is that we *must* do so (or adopt an equivalent description, such as percentages in which all the codes are artificially multiplied by 100). The *only* language of inference which deals consistently with simple problems is ordinary probability calculus, just as originally required by Laplace [4].

Any complicated proposition may be constructed from simpler ones, just as any complicated quantity in computer memory can be broken down into its constituent bits. Hence we must also use probability calculus when dealing with complicated problems. Indeed, the analogies with engineering and with computing suggest that strict submission to this discipline will become even more important as our problems become harder. Moreover, the analogies also suggest that the benefit of adhering to the discipline will be qualitative improvements in reliability, precision, and power, with benefits far outweighing the costs.

Those workers who do argue in strict probabilistic terms are called "Bayesians", although it seems perverse to have a special adjective (especially as Bayes' writings were less clear and extensive than those of Laplace). Outsiders are thereby given the impression that the strict probabilistic approach is just one of several, competing with other general schools such as fuzzy logic (Klir [5]), or with specific schools such as generalised cross-validation (Golub *et al.* [6]) and many more. As a matter of history, that has too often been the case, and it continues to the present day, but as a matter of philosophy it should not be. Consistent reasoning demands probability calculus. Logic has spoken.

We proceed to illustrate this with an analysis of the interpolation problem, which affords a particularly clean example of Bayesian methodology in action.

3   Interpolation Theory

The interpolation problem is to estimate a function $f(\mathbf{x})$ on the basis of a finite number N of data values $D_k = f(\mathbf{x}_k)$ at sample points $\mathbf{x}_k$ (*e.g.* Buhmann and Powell [7]). This is an inference problem. Without a great deal of extra information, we could not be sure that any particular f obeying the constraints was uniquely correct, so that we must seek the posterior probability $\Pr(f|D)$ as our inferred answer. In probabilistic language, the data give us

$$\Pr(D|f) = \prod_{k}^{N} \delta(f(\mathbf{x}_k) - D_k)$$

As usual, we need a prior $\Pr(f)$ which enables us to use the joint probability

$$\Pr(f, D) = \Pr(f)\Pr(D|f) = \Pr(D)\Pr(f|D)$$

to let us infer the numerical value of

$$\Pr(D) = \textbf{evidence}$$

as well as

$$\Pr(f|D) = \text{inference}$$

In order to interpolate successfully between the data points, our prior must exhibit some intrinsic spatial correlation. The natural way of encoding this (Gull [8]) is to suppose that f is a suitably blurred version of a "hidden" image h. Thus

$$f = Ch$$

where C is the intrinsic correlation matrix. The matrix C should be faithful to our underlying prior ideas about the configuration space $\mathbf{x}$ on which it operates. Being part of our prior model, it is to be independent of the particular points $\{\mathbf{x}_k\}$ which will (later) be chosen for measurement. Usually, the most natural assumption is that our knowledge in coordinate space is translation invariant, so that

$$C_{ij} \equiv C(\mathbf{x}_i, \mathbf{x}_j) = C(\mathbf{x}_i + \mathbf{y}, \mathbf{x}_j + \mathbf{y}) \qquad \forall \mathbf{y}$$

Accordingly, C will be a convolution operation

$$C_{ij} = \gamma(\mathbf{x}_i - \mathbf{x}_j)$$

which can, if necessary, be diagonalised by a Fourier transform.

Moreover, it is usually appropriate to assume rotation (or, in one dimension, inversion) invariance, in which case only the distance between two points is relevant. Thus

$$C_{ij} = \gamma(r_{ij}), \qquad r_{ij} = |\mathbf{x}_i - \mathbf{x}_j|$$

Variants of this in periodic or curved spaces are allowed, with the form of C being influenced by whatever group invariance is imposed.

Various forms for γ spring to mind, such as Gaussian, Lorentzian, square wave, *etc*, but there is a restriction. C should not have any zero eigenvalues. If C did have a zero eigenvalue, then f could not have any component along the corresponding eigenvector $\mathbf{e}$. We could check this by experiment, measuring $\mathbf{e}^T f$ directly. After all, C represents prior knowledge, and we could later decide to measure any feature of f which we wanted—we are not at this stage restricted to interpolation data. If we measured $\mathbf{e}^T f$ not zero, we would have a contradiction. Rather than lay ourselves open to this catastrophe, we shall legislate that C have no zero eigenvalues. If we wish to allow the eigenvalue spectrum of C to be continuous in the limit of continuous $\mathbf{x}$, then C must be one-signed. (It would be inelegantly artificial to suggest that C could change sign, but only discontinuously.) The natural choice of sign being positive, we require C to be a positive-definite matrix. Correspondingly, γ must have strictly positive Fourier transform, which allows a Gaussian or Lorentzian, but not a square wave.

Whichever form for γ is chosen, it should contain at least one numerical parameter to fix its scale. Such parameters and choices may well be influenced later by the form of the data, but for initial presentation we will allow C to be fixed.

The other part of our prior model is the hidden image h. By construction, this does not have any intrinsic prior correlation, because that is defined by C. Hence our prior $\Pr(h)$ factorises independently on the (arbitrarily large) number M of cells into which configuration space may be divided.

$$\Pr(h) = \prod_{i=1}^{M} \Pr(h_i)$$

The simplest choice is Gaussian, with

$$\Pr(h|\alpha) = \left(\frac{\alpha}{2\pi}\right)^M \exp\left(-\frac{\alpha}{2}h^T h\right), \qquad h^T h = \sum_{i=1}^{M} h_i^2$$

though another choice, appropriate for positive functions, would be

$$\Pr(h|\alpha) \;=\; Z(\alpha)^{-1}\exp(\alpha S(h))$$
$$S(h) \;=\; \text{entropy of } h, \qquad Z = \text{normalisation}$$

This latter choice leads to "maximum entropy" reconstructions. Because of the nonlinearity inherent in the entropy, the formalism is more complicated, in detail, but the numerical procedure parallel the Gaussian case closely. We proceed with the Gaussian prior bacause of its relative simplicity.

An extra parameter α, called the "regularisation constant", has been introduced into these priors, and explicitly incorporated into the notation $\Pr(h|\alpha)$. It effectively defines the units of h having (in the Gaussian case) dimensions $[h]^{-2}$. Hence it cannot be fixed to some *a priori* value.

Not knowing what α is, we must assign a prior to it. Presumably, we are initially ignorant of the units of h, within reasonably wide limits, and this leads us to place a correspondingly wide uniform prior on $\log \alpha$

$$\Pr(\log\alpha) = \begin{cases} 1/(\log\alpha_+ - \log\alpha_-) & \text{if } \alpha_- < \alpha < \alpha_+ \\ 0 & \text{otherwise} \end{cases}$$

We might appropriately think of $\alpha_- = 0.01, \alpha_+ = 100$. The details of this should not matter much. If they did, then our data would be unable to determine even the order of magnitude of the size of the reconstruction, and the dataset would be impoverished indeed. We will proceed on the basis of

$$\Pr(\alpha) = a/\alpha, \qquad a = 1/(\log\alpha_+ - \log\alpha_-) \sim 10$$

Our prior is now complete. We have, for initial presentation, fixed the correlation matrix C and the remaining quantities h and α have

$$\Pr(\alpha, h) = \Pr(\alpha)\Pr(h|\alpha) = \frac{a}{\alpha}\left(\frac{\alpha}{2\pi}\right)^{M/2}\exp\left(-\frac{\alpha}{2}h^T h\right)$$

With the prior fully defined, inference is henceforward automatic. All we have to do is follow the rules. Our full prior is

$$\Pr(\alpha, h, f|C) \;=\; \Pr(\alpha)\Pr(h|\alpha)\Pr(f|h, C)$$
$$=\; \frac{a}{\alpha}\left(\frac{\alpha}{2\pi}\right)^{M/2}\exp\left(-\frac{\alpha}{2}h^T h\right)\delta(f - Ch)$$

The data are incorporated to give the full joint probability distribution of everything relevant

$$\Pr(\alpha, h, f, D|C) \;=\; \Pr(\alpha, h, f|C)\Pr(D|f)$$
$$=\; \frac{a}{\alpha}\left(\frac{\alpha}{2\pi}\right)^{M/2}\exp\left(-\frac{\alpha}{2}h^T h\right)\delta(f - Ch)\delta(Pf - D)$$

where P is the $N \times M$ projection matrix from $\mathbf{x}$ to the N sample points $\{\mathbf{x}_k\}$. All subsequent inferences derive from this.

Our initial aim is to calculate the evidence $\Pr(D|C)$, which involves marginalising over α, h and f. We can do this in whichever order is convenient. Here, we choose to integrate out f first.

$$\Pr(\alpha, h, D|C) \;=\; \int d^M\!f\, \Pr(\alpha, h, f, D|C)$$
$$=\; \frac{a}{\alpha}\left(\frac{\alpha}{2\pi}\right)^{M/2}\exp\left(-\frac{\alpha}{2}h^T h\right)\delta(PCh - D)$$

Under the constraints $PCh = D$, the most probable hidden image is

$$\widehat{h} = C^T P^T A^{-1} D$$

Here, we have written

$$A = PCC^T P^T = \text{"dirty beam"}$$

A represents the feedback from one point to another, and plays a ubiquitous role in the analysis. Using $\widehat{h}$, we have the alternative expression

$$\Pr(\alpha, h, D|C) = \frac{a}{\alpha} \left(\frac{\alpha}{2\pi}\right)^{M/2} \exp\left(-\frac{\alpha}{2}\left(h - \widehat{h}\right)^T \left(h - \widehat{h}\right) - \frac{\alpha}{2} D^T A^{-1} D\right) \delta\left(PC\left(h - \widehat{h}\right)\right)$$

Next, we integrate out h. With h taking a Gaussian form centred on $\widehat{h}$, having N linear combinations fixed by the data, the integral is

$$\begin{aligned}\Pr(\alpha, D|C) &= \int d^M h \, \Pr(\alpha, h, D|C) \\ &= \frac{a}{\alpha} \left(\frac{\alpha}{2\pi}\right)^{N/2} (\det A)^{-1/2} \exp\left(-\frac{\alpha}{2} D^T A^{-1} D\right)\end{aligned}$$

Finally, we integrate out α

$$\begin{aligned}\Pr(D|C) &= \int d\alpha \, \Pr(\alpha, D|C) \\ &= a \, \Gamma\left(N/2\right) \left(\pi D^T A^{-1} D\right)^{-N/2} (\det A)^{-1/2}\end{aligned}$$

This central formula plays a crucial role in influencing our choice of intrinsic correlation matrix C. We may wish to choose different forms, and even within a given form there should always be at least one parameter to define the correlation scale. As Bayesians, we must formalise this by specifying some prior distribution $\Pr(C)$ over the matrices we are considering. From this we have the joint probability

$$\Pr(C, D) = \Pr(C)\Pr(D|C) = \Pr(D)\Pr(C|D)$$

Exactly as usual, we use this to find

$$\Pr(D) = \int dC \, \Pr(C, D) = \int dC \, \Pr(C)\Pr(D|C)$$

where $dC \equiv d(\text{parameters entering the matrix } C)$ and thence to estimate our choice of and our parameters for C

$$\Pr(C|D) = \Pr(C)\Pr(D|C)/\Pr(D)$$

It is for this reason that we call $\Pr(D|C)$ the **evidence**. With many data and few parameters, those few will usually be determined quite accurately. The high powers which appear in the expression for the evidence usually give a sharply peaked function of the parameters in C. Indeed, a special case of this appeared earlier, when the parameter α was distributed as $\alpha^{\frac{N}{2}-1}\exp(-\alpha K/2)$. Although we were then able to perform the α integral exactly as a gamma function, we could almost as well have just fixed α at the maximising value $\widehat{\alpha} = K/(N-2)$ or at the mean value $\widehat{\alpha} = K/N$. The prior on α influenced it merely by the single inverse power α^{-1}, which is quickly overwhelmed by the data. Similarly with C. Although the purist would integrate out all unknown parameters, it usually suffices to select average values, or even just those which maximise the evidence.

For convenience, we recommend that the evidence be quoted in decibels on a logarithmic scale as $10\log_{10}\Pr(D|\text{assumptions})$. Generally, the overall magnitude of the evidence $\Pr(D|\text{assumptions})$ powers with the number of data, as one might expect, so stray factors less than this are relatively unimportant. This means that we can, with reasonable equanimity, neglect probability factors less than about e^N in size!

We can thus use the evidence to fix our initially unknown choices and parameters. Even when this task is over, though, it is recommended practice to record the final numerical value of the evidence

$$\Pr(D) = \int dC \, \Pr(C) \, \Pr(D|C) \simeq \Pr(D|\hat{C})$$

Then, if someone else proposes a different prior model or prior structure, we can compare this with our recorded $\Pr(D)$, and see which is better, and by how much. This objective comparison of different theories is one of the major benefits of Bayesian analysis.

Now, having set C and calculated $\Pr(D|C)$, we reverse the calculation, and infer in reverse order all the parameters integrated out, as far as we need.

Top of the stack is α. From $\Pr(\alpha, D|C)$ and $\Pr(D|C)$ we have

$$
\begin{aligned}
\Pr(\alpha|D, C) &= \Pr(\alpha, D|C)/\Pr(D|C) \\
&= \frac{\alpha^{N/2-1}}{\Gamma(N/2)} \left(\frac{1}{2} D^T A^{-1} D\right)^{N/2} \exp\left(-\frac{\alpha}{2} D^T A^{-1} D\right)
\end{aligned}
$$

Provided N is not so small that we have almost no data, the mean value

$$\hat{\alpha} = N/D^T A^{-1} D$$

will adequately represent the distribution. Actually, even if N is quite small, little damage will ensue from thus fixing α.

Next is the hidden function h.

$$
\begin{aligned}
\Pr(h|D, C) &= \int d\alpha \, \Pr(h, \alpha|D, C) \\
&\simeq \Pr(h, \hat{\alpha}|D, C) \\
&= \Pr(h, \hat{\alpha}, D|C)/\Pr(D|C) \\
&= (2\pi)^{-M/2} (\det H)^{-1/2} \exp\left(-\frac{1}{2} \left(h - \hat{h}\right)^T H^{-1} \left(h - \hat{h}\right)\right)
\end{aligned}
$$

where

$$H = \hat{\alpha}^{-1} \left(I - C^T P^T A^{-1} P C\right)$$

is the covariance of h about its mean $\hat{h}$.

Lastly, we reach the reconstruction f in which we are primarily interested

$$
\begin{aligned}
\Pr(f|D, C) &= \int d^M h \, \Pr(f, h|D, C) \\
&= \int d^M h \, \Pr(h|D, C) \, \Pr(f|h, C) \\
&= (2\pi)^{-M/2} (\det Q)^{-1/2} \exp\left(-\frac{1}{2} \left(f - \hat{f}\right)^T Q^{-1} \left(f - \hat{f}\right)\right)
\end{aligned}
$$

where

$$Q = \hat{\alpha}^{-1} C \left(I - C^T P^T A^{-1} P C\right) C^T$$

is the covariance $\langle \delta f \delta f^T \rangle = Q$ of f about its mean

$$\hat{f} = CC^T P^T A^{-1} D$$

We can now make any deductions we want from f, obtaining not just a best estimate, but also probabilistic error bars. In particular, we can generate random samples

$$f = \widehat{f} + Q^{-1/2} g$$

g being a Gaussian random vector with components drawn from the unit normal distribution. All this can be accomplished equally well with the entropic prior, appropriate for postive f. In the Gaussian case, it happens that C only appears in the evidence, and in the inference about f, in the combination CC^T. In particular, the mean reconstruction $\widehat{f}$ is of the form

$$\widehat{f}(\mathbf{x}) = \sum_k^N c_k \phi(r_k)$$

where $r_k = |\mathbf{x} - \mathbf{x}_k|$ is the distance to the measured point k, c_k is the coefficient $(A^{-1}D)_k$. Also

$$\phi(r_{ij}) \equiv \phi(|\mathbf{x}_i - \mathbf{x}_j|) = \int d\mathbf{x}\, \gamma(\mathbf{x} - \mathbf{x}_i)\, \gamma(\mathbf{x} - \mathbf{x}_j)$$

is the component of CC^T evaluated at points r_{ij} apart, being the autocorrelation of γ. ϕ is known as a **"radial basis function"**, and $\widehat{f}$ appears as the sum of radial basis functions centred on the sample points.

The numerical analysis literature provides several examples of radial basis functions, and some discussion of which is "best". With our Bayesian viewpoint, we can decide objectively which is best for a given problem. However, must ensure that the matrix CC^T corresponding to ϕ is strictly positive, as befits a covariance matrix.

4   Choice of Radial Basis Function (RBF)

As already noted, we seek radial basis functions which have strictly positive Fourier transform. Where necessary, we make modifications to common functional forms in order to satisfy this condition. The following functions were used in our numerical investigations of interpolation in one spatial dimension.

4.1   Gaussian

This has the form

$$\phi(x) = \exp\left(-x^2/2x_0^2\right)$$

where x_0 is a length-scale or blur-width.

4.2   Periodic Spline

The simplistic cubic spline radial basis function $\theta(x) = |x|^3$ does not have positive-definite Fourier transform, so is not admissible as it stands. We modify it to the X-periodic form

$$\phi(x) = 3X^4 - 40y^2(X - y)^2 \qquad y = x \pmod{X}$$

in order to ensure Fourier positivity while retaining the third derivative discontinuity at the origin.

4.3   Truncated Multiquadric

The multiquadric

$$\theta(x) = \sqrt{x^2 + x_0^2} = |x| * \left(x^2 + x_0^2\right)^{-3/2}$$

where "*" denotes convolution, does not have a positive Fourier transform. However, the truncated form

$$\phi(x) = t(x) * \left(x^2 + x_0^2\right)^{-3/2}$$

where $t(x)$ is the "cusped triangle"

$$t(x) = \begin{cases} (w - |x|)^2 & \text{if } |x| < w \\ 0 & \text{otherwise} \end{cases}$$

does have a positive Fourier transform. This has much the same small-scale $\mathcal{O}(x_0)$ structure as θ while the large-scale $\mathcal{O}(w)$ structure has been modified appropriately.

4.4   Inverse Multiquadric

This has the form

$$\phi(x) = \frac{1}{\sqrt{x^2 + x_0^2}}$$

5   Results

We test each RBF on a smooth and a rough dataset each consisting of 26 equi-spaced datapoints. The interpolation is onto a grid in x which is sufficiently fine to pick up all the relevant structure. In each case, we select the value of x_0 (and w in the case of the truncated multiquadric) which maximises the evidence. The period X of the spline is fixed at a value suitably greater than the interpolation interval.

5.1   Smooth Data

This dataset is

$$x_i = 10i, \quad y_i = \sin\left(2\pi \left(\frac{i}{25}\right)^2\right) \qquad \text{for } i = 0, 1, \ldots 25$$

The following table shows the optimal parameters (in gridpoint units) and the corresponding evidence for each RBF:

RBF	$\widehat{x_0}$	$\widehat{w}$	Evidence
Gaussian	61	—	847 db
Spline	—	—	121 db
Truncated Multiquadric	275	50	779 db
Inverse Multiquadric	275	—	760 db

In fig. 1, the 26 datapoints are displayed as dots at locations $x_i = 10i$, $i = 0 \ldots 25$. The mean reconstruction $\widehat{f}$ is the solid line and the broken lines are error bars at the $\pm\sigma$ level. Since the results for the multiquadrics are visually identical to those of the Gaussian, we only display the Gaussian and spline results.

Because the data are smooth, the optimal blur-width is large in each case. At an evidence value of 847db, the Gaussian wins overall and the spline is poorest at 121db. The error bars within the data interval are so small that they are hardly discernable by eye though they increase in the extrapolated regions.

Fig. 2 shows the error bars on a logarithmic scale for each case. The error bar correctly falls to zero at the sample points. With the exception of the spline case, the reconstructions are more tightly constrained in the middle of the data interval than at the edges. This behaviour seems plausible—given smooth data on either side of a coordinate point, a correspondingly smooth datum at that point is not surprising; the datum is "predictable".

The size of the error bars gives an idea of the interval within which we expect typical reconstruction samples drawn from the posterior distribution to vary. The smoothness of such samples will depend on the RBF used and will not necessarily mimic the smoothness of the average sample, $\hat{f}$. We shall return to the subject of sampling from the posterior.

5.2 Rough Data

These data are

$$x_i = 10i, \quad y_i = 0 \quad \text{for } i = 0, 1, \ldots 6 \text{ and } 19, 20, \ldots 25$$
$$x_i = 10i, \quad y_i = 1 \quad \text{for } i = 7, 8, \ldots 18$$

The optimal parameters and evidence values are as follows:

RBF	$\hat{x_0}$	$\hat{w}$	Evidence
Gaussian	9.5	—	-35.7 db
Spline	—	—	-83.8 db
Truncated Multiquadric	1	95	-13.5 db
Inverse Multiquadric	12.5	—	-20.1 db

The interpretation of the plots of fig. 3 is as in fig. 1. The optimal blur-width values are now much smaller to accommodate the sharp step in the data. Because of this step, the error bars within the data interval are now comparatively large. In particular, for the truncated multiquadric case, which wins overall on this dataset, the error bars are large enough for the linear interpolant to fall entirely within them. Indeed, the value $x_0 = 1$ is so small that the mean interpolant is nearly a linear interpolant.

Fig. 4 shows the corresponding error bars as in fig. 2, but on a linear scale.

5.3 Sampling from the Posterior

We have seen that the probabilistic analysis provides more than a single reconstruction; the inference about f is the posterior distribution $\Pr(f|D)$ from which we obtain the mean reconstruction $\hat{f}$ and error bars. We may also sample from this distribution to obtain representative reconstructions. Displaying such a family of samples is more informative about the robustness of interpolated features than the mere display of the mean and error bars. We can, indeed, display a smoothly continuous "movie" by randomly diffusing through the posterior and displaying samples along the diffusion path. We recommend this excellent visual representation, but for this paper we can only display a characteristic sample.

The case of rough data is particularly interesting. The Gaussian RBF used in fig. 3(a) is the autocorrelation of the ICF (also a Gaussian) shown in fig. 5(a). The correlation length-scale of an interpolant is given by the width of this ICF (we recall that the interpolant is a convolution of the ICF with a hidden uncorrelated image). Thus we should expect a sample from the posterior to exhibit correlation of the same length-scale. This is borne out by the sample shown in fig. 5(b), which plausibly lies largely within the error bars and oscillates on a length-scale comparable to the width of the ICF.

The ICF of the truncated multiquadric of fig. 3(c) is shown in fig. 5(c). It is much sharper an narrower than the Gaussian ICF in the central region but it has broader wings. As a result, it ca accommodate both sharp and smooth structure in the data to produce an interpolant with ver little "Gibbs ringing" as seen in fig. 3(c). However, a typical sample from the posterior (fig. 5(d) oscillates rapidly—in keeping with the very small length-scale of the ICF. The mean interpolant which is much smoother, is a very special sample from the posterior for which the small scal oscillations average out.

6 Conclusion

All scientific inference ought to be cast in the language of probability theory. One's conclusion abou any quantity, whether it be a simple scalar or a complicated function, is a probability distributio for that quantity. This rests upon a prior distribution, for which there may be a wide choice However, any such choices can be objectively assessed using the probability formulae themselves We have illustrated such assessment with interpolation data, which are particularly easy to visualis and discuss.

Such data require us to introduce explicit local correlation into the prior, and we do this througl an intrinsic correlation function. The results in this paper were obtained from an underlying Gaussian prior, for simplicity of exposition. However, the formulae generalise naturally to a underlying entropy prior, such as has been used to date in maximum entropy analysis.

We expect this incorporation of local correlation to yield significant improvements in all form of maximum entropy reconstruction.

References

[1] Skilling, J. (1989) "Classic Maximum Entropy", in *Maximum Entropy and Bayesian Methods* ed. Skilling J., Kluwer, 45-52.

[2] Cox R.P. (1946) "Probability, Frequency and Reasonable Expectation", *Am. Jour. Phys.* **17** 1-13.

[3] Jaynes E.T. (1989) "Clearing up Mysteries—the Original Goal", in *Maximum Entropy and Bayesian Methods* ed. Skilling J., Kluwer, 1-27.

[4] Laplace, P.S. (1814) *Essai Philosophique sur les Probabilités*, Courcier Imprimeur, Paris.

[5] Klir, G. J. (1987) "Where Do We Stand on Matters of Uncertainty, Ambiguity, Fuzziness and the Like?", *Fuzzy Sets and Systems* **24**, 141-160.

[6] Golub, G., Heath, M., Wahba, G. (1979) "Generalised Cross Validation as a Method for Choosing a Good Rdige Parameter", *Technometrics* **21**, 215-224.

[7] Buhmann, M.D., Powell, M.J.D. (1990) "Radial Basis Function Interpolation on an Infinite Regular Grid", in *Algorithms for Approximation II*, eds. Mason, J.C., Cox, M.G., Chapman and Hall, pp. 146-169.

[8] Gull, S.F. (1989) "Developments in Maximum Entropy Data Analysis", in *Maximum Entropy and Bayesian Methods* ed. Skilling J., Kluwer, 53-71.

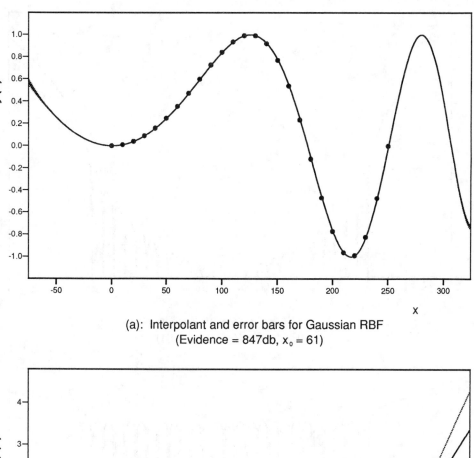

(a): Interpolant and error bars for Gaussian RBF
(Evidence = 847db, $x_0 = 61$)

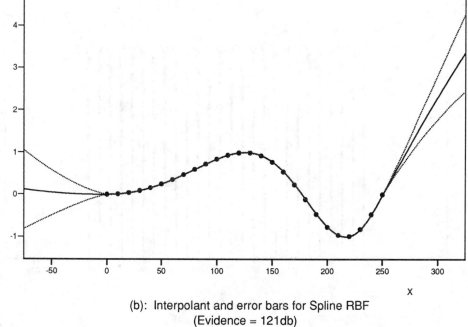

(b): Interpolant and error bars for Spline RBF
(Evidence = 121db)

Fig 1: BAYESIAN INTERPOLATION OF SMOOTH DATA

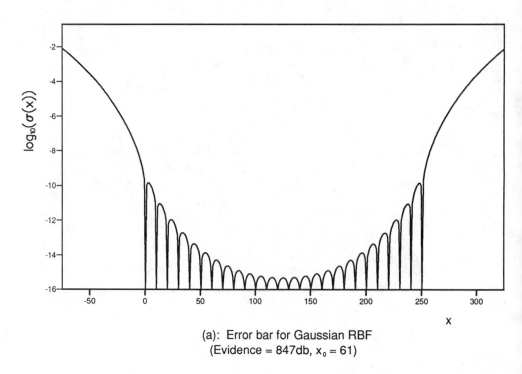

(a): Error bar for Gaussian RBF
(Evidence = 847db, $x_0 = 61$)

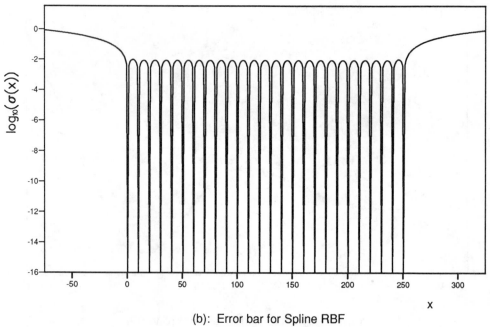

(b): Error bar for Spline RBF
(Evidence = 121db)

Fig 2: BAYESIAN INTERPOLATION OF SMOOTH DATA

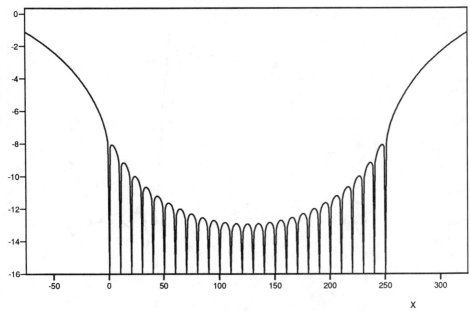

(c): Error bar for Truncated Multiquadric RBF
(Evidence = 779db, $x_0 = 275$, w = 50)

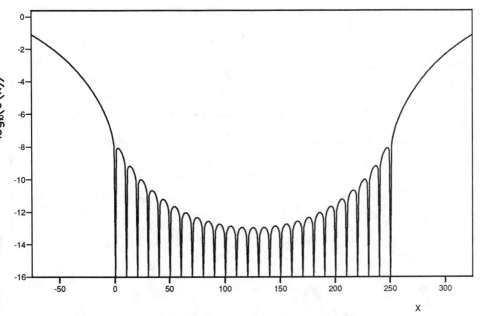

(d): Error bar for Inverse Multiquadric RBF
(Evidence = 760db, $x_0 = 275$)

Fig 2: BAYESIAN INTERPOLATION OF SMOOTH DATA

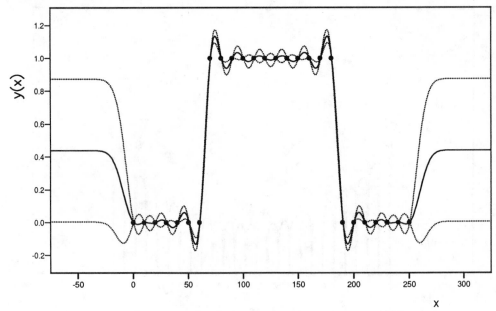

(a): Interpolant and error bars for Gaussian RBF
(Evidence = -35.7db, x_0 = 9.5)

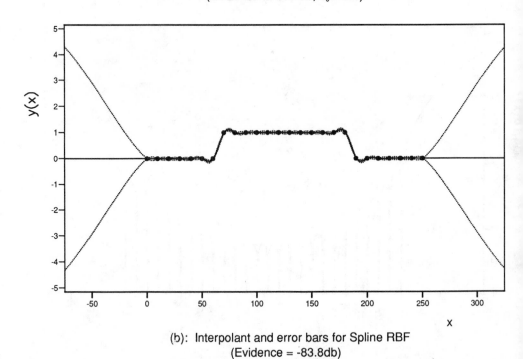

(b): Interpolant and error bars for Spline RBF
(Evidence = -83.8db)

Fig 3: BAYESIAN INTERPOLATION OF ROUGH DATA

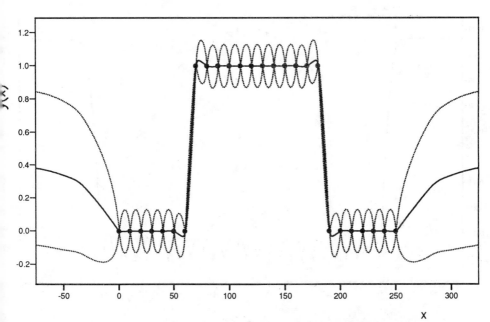

(c): Interpolant and error bars for Truncated Multiquadric RBF
(Evidence = -13.5db, $x_0 = 1$, $w = 95$)

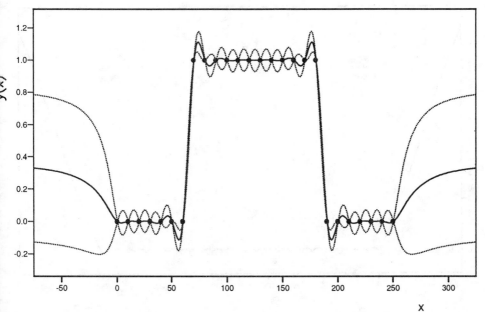

(d): Interpolant and error bars for Inverse Multiquadric RBF
(Evidence = -20.1db, $x_0 = 12.5$)

Fig 3: BAYESIAN INTERPOLATION OF ROUGH DATA

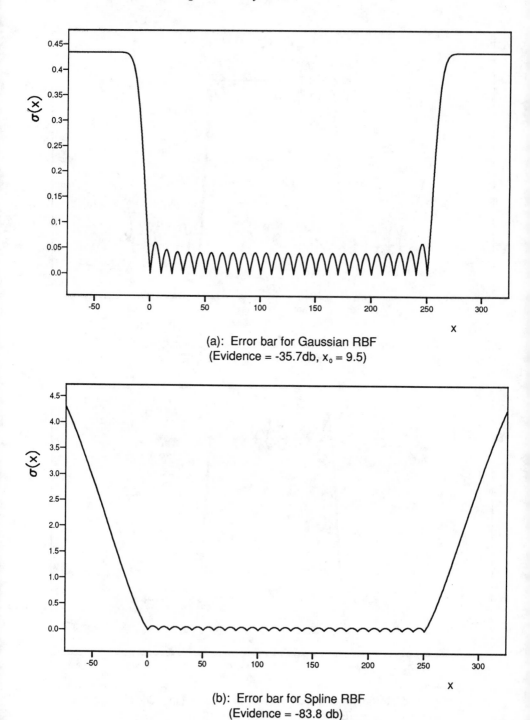

(a): Error bar for Gaussian RBF
(Evidence = -35.7db, x_0 = 9.5)

(b): Error bar for Spline RBF
(Evidence = -83.8 db)

Fig 4: BAYESIAN INTERPOLATION OF ROUGH DATA

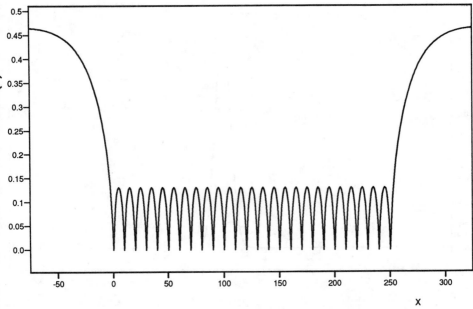

(c): Error bar for Truncated Multiquadric RBF
(Evidence = -13.5db, $x_0 = 1$, $w = 95$)

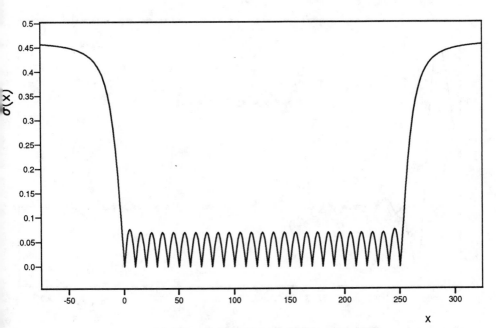

(d): Error bar for Inverse Multiquadric RBF
(Evidence = -20.1db, $x_0 = 12.5$)

Fig 4: BAYESIAN INTERPOLATION OF ROUGH DATA

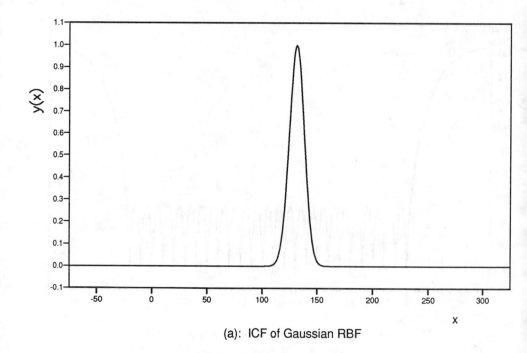

(a): ICF of Gaussian RBF

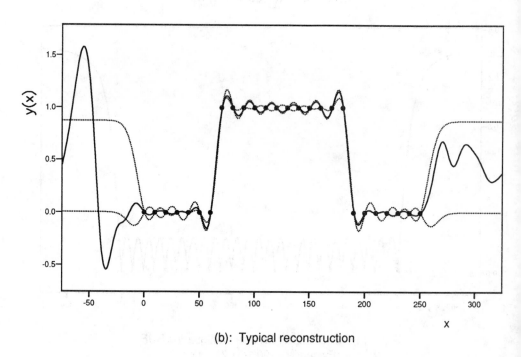

(b): Typical reconstruction

Fig 5: SAMPLES FROM POSTERIOR FOR ROUGH DATA

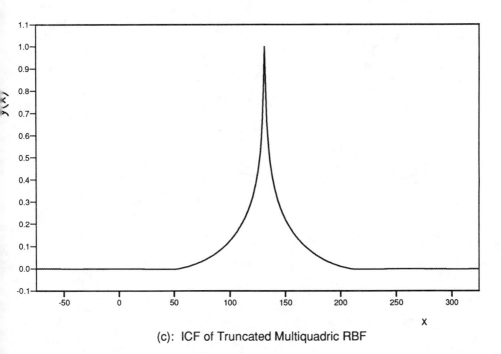

(c): ICF of Truncated Multiquadric RBF

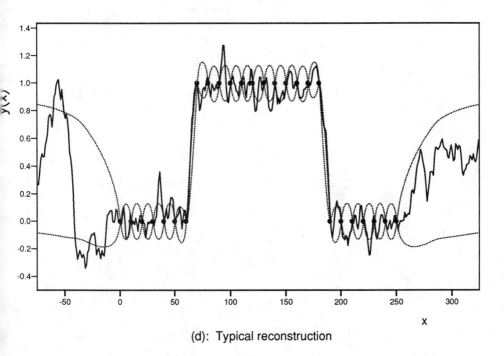

(d): Typical reconstruction

Fig 5: SAMPLES FROM POSTERIOR FOR ROUGH DATA

Inst. Phys. Conf. Ser. No 107: Chapter 1
Paper presented at Neutron Scatt. Data Anal. Conference, Rutherford Appleton, 1990

23

Simulated annealing: an introductory review

J Pannetier

Institut Laue-Langevin, 156X - 38042 Grenoble Cedex (France)

ABSTRACT: Simulated annealing is a global optimization procedure which is based on an analogy with the physical process of annealing. This article gives an introduction to the concepts and technicalities of the simulated annealing algorithm and describes some of its applications in condensed matter physics. It is concluded by a classified bibliography on the algorithm and its applications.

1. INTRODUCTION

Simulated annealing (SA) is a general purpose *global* optimization technique for very large combinatorial problems which was introduced in 1983 by S. Kirkpatrick, C.D. Gelatt and M.P. Vecchi in an article in Science [1.1]. Over a few years, interest for SA turned to enthousiasm as it proceeded to provide improved solutions to many optimization problems in many fields. To date most of the development and applications of SA have been in the area of large-scale combinatorial optimization (problems with discrete variable parameters) but it applies as well to optimization over continuous variables. Curiously enough, SA has been applied to a wide range of engineering optimization problems in such diverse areas as image processing, circuit placement and wiring in computer-aided electronic design, scheduling problems, pollution control and even deployment of missile interceptors but, although it originates from statistical mechanics, it is still largely overlooked in condensed matter research. The scope of this article is two-fold : first provide an introduction to SA by presenting its origin and basic concepts and giving a detailed description of the implementation of the SA algorithm for practical applications; second illustrate some of its applications with special emphasis on problems of crystal and molecular structure determination and refinement.

2. ORIGIN AND BASIC CONCEPTS OF SIMULATED ANNEALING

Problems of optimization (function minimization) can be divided in two general classes :
- local optimization, in which one attempts to find the minimum of a function by exploring the neighbourhood of a starting solution (iterative improvement method)
- global optimization, in which one searches over all the configuration space of the problem to locate the global minimum.
The first class of optimization methods is more easily amenable to theoretical analysis and performs well for unimodal functions; however, in the case of multimodal functions, these methods tend to get trapped in the first minimum encountered and this local minimum will depend on the initial configuration (starting point of the optimization). A possible method to overcome the limitations of local search algorithms is to execute the search for a large number of different initial solutions and accept the best solution as the answer to the problem to solve;

References noted [*.*] refer to the selected list of papers on SA given in appendix; references quoted by a single number in parentheses [*] are given at the end of the text. For sake of convenience, the references of § 5 and 6 are given at the end of the corresponding section.

although such an approach would asymptotically find a global optimum with probability one, it would increase considerably the computation time; for instance, in the case of combinatorial problems the number of possible configurations is factorially large (see e.g. the travelling salesman problem - TSP - in § 5.1). The fundamental limitation of iterative improvement methods arises obviously from the fact that during the exploration of the solution space only solutions that decrease the cost function are accepted (hence their name of downhill algorithms). An alternative approach would be to accept in a limited way transitions which correspond to an increase in cost in order to escape from local traps of the energy landscape. This is exactly what the simulated annealing algorithm does; its name (it is also known, among other names, as statistical cooling and probabilitic hill climbing algorithm) refers to the analogy with the annealing of solids. Indeed the basic idea behind the simulated annealing originates from statistical mechanics. It is based on the observation [1.1] that combinatorial optimization problems with a large space exhibit properties similar to physical processes with many degrees of freedom : bringing a sample to a highly ordered crystalline state of minimum energy (e.g. growing a single crystal) requires a careful annealing, first melting the substance and then lowering the temperature slowly, especially in the vicinity of the crystallization temperature; the reduction of the temperature confines the system to a smaller and smaller region of the configuration space but, if carried out slowly enough, allows the system to escape from metastable local energy minima. Starting off from a formal equivalence between the many parameters in an optimization problem and the particles in a physical system and identifying the cost in the problem with the energy of the physical system, Kirkpatrick [1.1] proposed to apply to optimization problems the computer simulation methods used in condensed matter physics. In this respect, it is worth to emphasize the importance of the concept of frustration in optimization problems. As in spin-glasses, most difficult optimization problems involve competition between different and often incompatible kinds of ordering which induces degeneracy in the low-energy states of the problem and yields complex energy landscapes.

In application to combinatorial optimization, the method is just iterative improvement done at a sequence of finite "temperatures", the temperature being in this context a control parameter. The process starts at "high temperature" and the Metropolis algorithm [1] is used to explore the configuration space through random walks : starting from an arbitrary configuration ω with "energy" $E(\omega)$, one applies a small perturbation to produce a candidate new configuration ω' with energy $E(\omega')$. The transition probability for the acceptance of the new configuration is given by the Boltzmann statistics :

$$P(\omega \rightarrow \omega') = \min\left[1, \exp\left(\frac{-(E(\omega') - E(\omega))}{kT} \right) \right]$$

After a sufficient number of configurations have been sampled, the average value of the energy stabilizes at a value appropriate to this temperature. The temperature is then lowered according to some "annealing schedule", the system re-equilibrated and so on until the equilibrium energy does not change significantly on lowering the temperature further. This technique can therefore be regarded as a Monte Carlo (MC) iterative improvement method. The use of a Boltzmann probability distribution of the energy implies that the system sometimes goes uphill as well as downhill : this occurs frequently at high temperature but becomes less likely at lower temperatures.

3. SIMULATED ANNEALING AT WORK

From a practical point of view the application of any iterative improvement algorithm (including simulated annealing) requires the definition of the following elements : a concise description of the system being optimized, a cost function and a generation mechanism; in addition to that, SA requires to define an adequate annealing schedule.

3.1 Configuration of the system to optimize

This first step, common to all optimization problems, simply consists in giving a description of possible system configurations (solutions) by identifying the variable parameters of the problem to solve. This description will of course depend on the system to be investigated. For instance, in the case of a real physical system one will use a set of atomic positions or spin orientations. For combinatorial problems it could be the order of the cities (TSP), a set of locations (placement problem), etc.

3.2 Cost function (energy) E

The cost function (sometimes termed score) which is the equivalent of the energy of a physical system is a real number assigned to each possible configuration of the system. It represent the cost effectiveness of a solution with respect to the quantity that is to be optimized. Its definition obviously depends on the description of the system : for the TSP it is the tour length whereas for a problem of crystal structure determination, it would be the R-value or any related agreement factor. The definition of a proper cost function E which is an essential step of the optimization may actually be less than obvious for most engineering applications of SA optimization. In particular, it may imply to select proper weights to scale different, and often contradictive, terms in the total cost function.

3.3 Generation of local arrangements

In the original formulation of the simulated annealing algorithm the method used to generate new configurations is the Metropolis importance sampling procedure [1]. This technique usually known as the Monte-Carlo method [2] is widely used to estimate the equilibrium structural and thermodynamic properties of many-particle systems interacting with a specified Hamiltonian. Applied to a physical system of N particles confined in a volume V at temperature T (canonical ensemble), the procedure consists in the generation of a sequence of states, often called a Markov chain, with constant transition probabilities. The states of the chain are points in the D.N dimensional configuration space of the system of dimension D. The object of the method is to generate a Markov chain in which asymptotically each state ω recurs with a frequency proportional to its Boltzmann factor $\exp[-E(\omega)/kT]$; that is, instead of choosing configurations randomly then weighting them with the Boltzmann factor (which is computationally very inefficient), one selects configurations with a probability $\exp[-E(\omega)/kT]$ and weight them evenly. The equilibrium value of any state function Φ of the configuration state will be the asymptotic limit of the average over the chain of that quantity. The convergence is independent of the initial state. Technically, the Metropolis algorithm is a realisation of a transition matrix $P(\omega,\omega')$ which gives the conditional probability of a transition from a state ω to a state ω' at one step of a Markov chain. This transition matrix must satisfy the conditions :

$$P(\omega,\omega) = 1 - \sum_{\omega \neq \omega'} P(\omega,\omega')$$

$$P(\omega,\omega').\exp\left[\frac{-E(\omega)}{kT}\right] = P(\omega',\omega).\exp\left[\frac{-E(\omega')}{kT}\right]$$

where $E(\omega)$ is the energy of configuration ω. The first condition springs from the fact that ω' includes all possible configurations accessible from ω (normalization) and the second is the statement of detailed balance between ω and ω' (microscopic reversibility). The one-step transition probabilities are determined by the product of the probability g of generating a given configuration and the probability f of accepting it. For numerical simulation of a physical system, the function $g_i(\omega,\omega')$ describes the method of choosing new positions for particle i whereas $f_i(\omega,\omega')$ describes the criterion for determining whether or not to accept the new position. The transition probabilities can thus be written as :

$P(\omega,\omega') = g_i (\omega,\omega'). \min[1, f_i (\omega,\omega')]$. For the standard Metropolis scheme, these functions are chosen as :

$$g_i(\omega,\omega') = \begin{cases} 1 \text{ if } | x_i(\omega') - x_i(\omega) | < \delta \\ \\ 0 \text{ otherwise} \end{cases} \quad \text{and} \quad f_i(\omega,\omega') = \exp -\left[\frac{E(\omega') - E(\omega)}{kT} \right]$$

where δ denotes the radius of the sphere within which the new position for article i is chosen. The same acceptance probability f can be used to generate sequences of configuration for an optimization problem : the cost function will simply take the role of energy while the quantity kT in the Boltzmann term is usually called the control parameter. For combinatorial problems (discrete configuration space) the generation mechanism g will usually be based on an exchange mechanism (e.g. for TSP, reverse the order in which an arbitrary sequence of cities in the current tour are traversed; for a placement problem, select two cells at random and exchange their position).

3.4 Annealing schedule

In the SA procedure temperature is considered as a control parameter : by slowly decreasing this parameter from a high value ("molten" state), the system will hopefully be frozen in a state of minimum energy. The search for adequate annealing schedule, crucial to reach efficiently a near-global optimum, has been addressed in many recent papers (for a detailed mathematical analysis of the algorithm the reader is urged to consult [2.1] and references quoted therein) but most of the schedules mentionned in the litterature remain largely based on empirical rather than theoretical rules. Nevertheless all recent theoretical studies of the algorithm agree on the conclusion that the speed of the algorithm and the quality of the final configuration are governed essentially by four parameters :

3.4.1 *Initial value of the control parameter (temperature)* :

It is intimately related to the average value of the cost function and has to be large enough to ensure that initially all configurations of the system occur with equal probability. The choice of the optimal initial temperature T_0 is still subject to some discussions in the litterature and theoretical estimates are often too large for computational efficiency. Practically it must be chosen in such a way that almost all proposed moves are accepted, i.e. such that $\exp(-\Delta E/T_0) \approx$ 1. A conceptually simple approach to estimate T_0 is Kirkpatrick's trial-and-error method : starting from an arbitrary temperature, attempt a few hundred moves and determine the fraction A of accepted moves. If that fraction is less than a given value A_0 (often taken as 0.8), the temperature is doubled; the procedure is carried again until the acceptance ratio A exceeds the threshold and the final value of temperature taken as T_0. In the physical analogy this corresponds to heating up a solid until it melts. More elaborate methods to estimate T_0 have been proposed by a number of authors; they usually consist in monitoring the evolution of the system during a number of rearrangements of the system before starting the actual optimization process :

• Aarts and van Laarhoven [2.1] recommend that T_0 is determined by increasing the value of T until a constant value of the acceptance ratio A is maintained. The value of T_0 is then given by the expression :

$$T_0 = \frac{<\Delta E^+>}{\ln (A^{-1})}$$

where $<\Delta E^+>$ is the average of the ΔE values for which $\Delta E > 0$ (uphill transitions).

• White [5.1] has proposed to estimate T_0 by the following method (hot condition) : using a high temperature for which all transitions are accepted ($T_\infty <-> A \approx 1$), perform a number of random transitions to calculate the averages $<E>$ and $<E^2>$. The initial temperature is then calculated as :

$$T_0 = k.(<E_\infty^2> - <E_\infty>^2)^{1/2}$$

where k is computed [5.7] assuming a normal cost distribution at high temperature. If T_0 is chosen such that a configuration whose cost is 3σ worse than the average one is accepted with probability P then $k = -3 / LnP$; a typical value of k is ≈ 20. White has also remarked that some systems may exhibit different classes of moves and thereby different characteristic temperature scales; such systems require appropriate cooling schedules.

Finally it should be noted that setting $T_0 = 0$ reduces the annealing algorithm to a simple iterative improvement method ("greedy" algorithm) since the acceptance criterion then allows only new configurations which decrease the cost function.

3.4.2 *Length of the Markov chain.*

The number L_t of MC cycles to perform at each temperature is determined by the concept of quasi-stationarity but is notoriously difficult to calculate and its estimation is still semi-empirical. The simplest choice is a value depending polynomially on the size of the problem; L_t is then independent of temperature. However this choice may not be computationally very effective; indeed the analogy with physical systems suggests that equilibrium is reached rapidly at high temperature but requires a longer time (larger number of transitions) at lower temperature. Starting from this observation more sophisticated methods have been formalized :

• a simple one is based on the intuitive argument that, for each temperature, a minimum number of transitions should be accepted. To avoid exceedingly long Markov chains at low temperature, when the acceptance rate drops down, L_t must however be ceiled by some constant L_{max}, usually chosen as a multiple of the number of variable of the problem.

• Romeo et al. [5.3] suggested that, at any temperature during the annealing process, there should be a sufficiently large probability to jump out of any local minimum. By developing this argument, they proposed to estimate L_t as :

$$L_t \propto \exp\left[\frac{-(E_{max} - E_{min})}{T_t}\right]^{-1}$$

where E_{max} and E_{min} are the highest and lowest values of the cost function obtained so far at this temperature during the execution of the algorithm.

• Huang *et al.* [5.7] based their estimation of L_t on the observation that, once equilibrium is reached, the ratio of the number N_{within} of new states generated with their energy E within a certain range δ from the average energy : $<E> - \delta < E < <E> + \delta$ to the total number N_{total} of accepted states also reaches a stationary value χ. Assuming a normal distribution of the values of energy , Huang *et al.* have shown that $\chi = erf(\delta/\sigma)$ where σ is the standard deviation of the energy distribution. Equilibrium can be monitored by comparing the "within count" M_{within} and the "outside count" $M_{outside}$ with given target values T_{within} and $T_{outside}$. Using a typical value for δ of $0.5*\sigma$ which yields $\chi = erf(0.5) = 0.38$, the target values are arbitrarily set to : $T_{within} = 0.38*[3*size of the problem]$ and $T_{outside} = 0.62*[3*size of the problem]$. Equilibrium is considered established if M_{within} reaches T_{within} before the outside count $M_{outside}$ reaches exceeds the maximum limit $T_{outside}$. If the tolerance limit $T_{outside}$ is exceeded, both M_{within} and $M_{outside}$ are reset to zero and the counting is initiated again. Some additional conditions must be imposed to avoid extremely long Markov chains at low temperatures because at these temperatures virtually no configurations are accepted and the "within count" M_{within} may never reach the target value.

It is worth mentioning that the length L_t of the Markov chain and the cooling program (see § 3.4.3) are strongly related : large temperature decrements will require longer Markow chains to re-establish equilibrium of the system.

3.4.3 *Cooling program.*

It consists in a sequence of monotonically decreasing temperatures. That decrease must be as fast as possible ("aggressive cooling") to be computing cost efficient but slow enough to maintain the system in quasi-equilibrium (avoid quenching) and increase the chances of obtaining good solutions. The most common cooling law uses the geometric decrement function first proposed by Kirkpatrick [1.1] :

$$T_t = \alpha . T_{t-1} = \alpha^t . T_0 \qquad t = 1, 2, \ldots$$

where α is a constant usually chosen in the range $0.8 \leq \alpha < 1$ and t denotes the successive (homogeneous) Markov chains generated at constant temperature T_t. Although it performs reasonably well in practice, more elaborate decrement rules based on an analysis of the parameters that govern the convergence of the algorithm have also been proposed :

• Geman and Geman [5.2] in their analysis of "stochastic relaxation" have shown that if the temperature T_t employed satisfies the bound :

$$T_t \geq \frac{c}{\log(1+t)} \qquad t = 0,1,2,\ldots\ldots$$

for every t, where c is a constant independent of t, then the configurations generated by the algorithm will tend to those of minimal energy as t -> ∞ ; unfortunately, this result provides only an upper limit for the cooling rate and the best theoretical values of c are far too large for computational purposes. The problem was further investigated by Mitra *et al.* [5.5] on the basis of the theory of time-inhomogeneous Markov chains; they proved that for update functions of the form :

$$T_t = \frac{\gamma}{\log(t+t_0+1)} \qquad t = 0,1,2,\ldots$$

where t_0 is any parameter satisfying $1 \leq t_0 \leq \infty$, the Markov chain is strongly ergodic (i.e. any point in the configuration space may be reached) if $\gamma \geq rL$ where r is a parameter which characterizes the connectivity of the graph underlying the chain and L a constant bounding the local slope of the cost function. Unfortunately estimates of γ can be obtained only for some combinatorial optimization problems and the above expression remains of little practical use.

• According to Aarts and van Laarhoven [2.1], the annealing algorithm can be optimised by choosing the temperature decrement and the Markov-chain length such that the process stays in quasi-equilibrium. This requirement can be achieved by using a cooling function of the form :

$$T_t = T_{t-1}\left[1 + \frac{\ln(1+\delta)T_{t-1}}{3\sigma(T_{t-1})} \right]^{-1}$$

where δ is a small positive number called the distance parameter and $\sigma(T_{t-1})$ denotes the standard deviation of the values of the cost function of the configurations of the Markov chain at temperature T_{t-1}; it can be calculated, as the algorithm proceeds, from the variance of the cost function :

$$\sigma^2(T_{t-1}) = <E^2(T_{t-1})> - <E(T_{t-1})>^2$$

The deviation from the optimum of the final value of the cost function increases when δ becomes larger. Typical values for δ are in the range 0.01 to 1.0.

• Huang *et al.* [5.7] use a curve of average cost versus the logarithm of temperature to guide the temperature decrease. To maintain quasi-equilibrium, they require that the difference in average energy for two consecutive temperatures T_{t-1} and T_t is less than the standard deviation of the energy, i.e. $E(T_t) - E(T_{t-1}) = -\lambda.\sigma(T_{t-1})$ where $\lambda \leq 1$. This results in cooling law of the form :

$$T_t = T_{t-1} . \exp\left[\frac{-\lambda T_{t-1}}{\sigma(T_{t-1})} \right]$$

A typical value of λ for practical applications is 0.7.

• J.D. Nulton and P.Salamon [5.11] recently presented a different theoretical approach by relating the problem of choosing an annealing schedule to that of determining an appropriate measure of distance in probability space. Defining the "target" (at a given temperature) as the equilibrium distribution towards which the system would relax asymptotically if the value of the temperature were held fixed, this distance quantifies the separation between the system and the target as the measure of statistical uncertainty associated with their distinguishability. The proposed theoretical criterion for the design of a temperature schedule takes the simplified form:

$$\frac{dT}{dt} = \frac{-vT}{\varepsilon\sqrt{C}}$$

Its implementation requires a knowledge of the heat capacity C of the target and the relaxation time ε of the system; v which is called the thermodynamic speed of the process is equal to the thermodynamic distance between the system and the target. Application of this criterion requires the evaluation of C and ε and appears to be rather difficult for most combinatorial problems; a method is outlined in [5.11] for calculating these quantities from first principles but only preliminary results are available yet.

• The aforementionned schedules correspond to the standard simulated annealing method used in most optimization problems. In the generalized simulating annealing (GSA) method introduced by I.O. Bohachevsky *et al.* [4.3], the "temperature" of the sample is held fixed at some arbitrary value, but the function f used to accept or refuse a new configurations is altered to include a function of the objective function. Assuming that the objective function equals zero at the global extremum, the acceptance function is given by :

$$f_i(\omega,\omega') = \exp - \left[E^m(\omega)\frac{E(\omega') - E(\omega)}{kT} \right]$$

where m is an arbitrary negative number. For m = 0, one recovers the standard simulated annealing procedure. The restriction of a zero global extremum can obviously be removed by introducing the value E_{min} of the global minimum and substituting $(E - E_{min})$ to E in the previous relation. A difficulty of the method remains however for most applications : the energy of the minimum is generally not known ! The difficulty can be overcome by guessing an initial value of E_{min}, running this algorithm until the difference $(E - E_{min})$ becomes negative, then guessing a new value for E_{min} and so on. However this procedure must be adapted for each particular problem; the method is therefore not of general applicability and its ability to reach a global optimum remains to be proved. Evaluation of GSA on a set of standard test problems [4.7] has shown that it is usually competitive with other methods; however in application to functions with an unknown global minimum value, the performance of the algorithm was found to be rather sensitive to the estimated minimum E_{min}.

3.4.4 *Stopping criterion.*

It is also known as the frozen condition and determines the final value of the temperature. Termination of the algorithm often uses one of the following simple criteria :

• the total number of MC cycles for which the algorithm has been executed is larger than a preset value : this criterion limits the computing cost but obviously does not imply that near-optimum convergence is reached.

• the acceptance ratio A becomes smaller than a given value A_f; this indicates that the system is effectively frozen but does not mean that the algorithm has converged.

More advanced criteria are based on a monitoring of the energy during annealing :

• Huang *et al.* [5.7] method is based on the following quantities calculated at the end of each Markov chain (i.e. before decreasing temperature) :
- the difference between the maximum and minimum values of the cost function in the chain

$$\delta_1 = E_{max} - E_{min}$$

- the maximum change in cost for any transition accepted during the generation of that chain :

$$\delta_2 = \max[E_t - E_{t-1}]$$

If they are the same, all the configurations in the Markov chain are of comparable energy and there is no need to use further simulated annealing.

• a related method was proposed earlier on by Vanderbilt and Louie [4.1] for continuous variable problems and is based on the calculation of the ratio $[<E> - E_{min}] / <E>$ where E_{min} is the minimum value of the cost function observed in the Markov chain. Typically, the algorithm is stopped when this ratio falls below an arbitrary value (e.g. 10^{-3}).

It is worth to note that in the last stages of the SA optimization procedure (namely when the system configuration is in the basin of attraction of the solution) is is advisable to switch to a more efficient method of optimization (greedy algorithm).

3.5 Monitoring the behavior of the algorithm

While the SA algorithm can in principle achieve convergence to a globally-optimum solution with probability arbitrarily close to 1 (given enough computer time), for practical applications the result strongly depends on the parameters of the annealing. Therefore several tools (annealing curves) have been developed to assess the progress of the procedure and judge the quality of the convergence of the SA algorithm.

3.5.1 *Energy curves E(T) and <E(T)>*

For discrete problems Energy vs. Temperature plots often show similar curves which, in the high temperature regime, can be approximated [5.1] as $<E(T)> \approx <E(\infty)> - \sigma^2(T) / T$ where $\sigma^2(T) = <E^2(T)> - <E(T)>^2$ is the variance of the cost function in equilibrium. The qualitative behavior of $<E>$ at lower temperature, in the case of discrete problems, has been discussed by White [5.1] and Van Laarhoven & Aarts [2.2].

3.5.2 *Variance $\sigma^2(T)$ and specific heat*

As occurs during the annealing of physical systems to obtain ordered materials, a uniform temperature reduction over the complete range of temperature may not be the most efficient procedure. There is often a crucial temperature range (e.g. close to a crystallization temperature) in which slow cooling is required whereas the temperature decrement can be much faster above and below this range. By analogy with the cooling of a fluid, this transition region can be identified by monitoring the temperature evolution of the fluctuations of the cost function, either by its variance $\sigma^2(T)$ or, in mechanical statistics terms, through the specific heat :

$$C(T) = \frac{d<E(T).}{dT} = \frac{\sigma^2(T)}{kT^2}$$

A large peak in $C(T)$ signals a change in the state of order of a system [3.3], i.e., in optimization, means that the configuration is becoming frozen into a energy minimum. Curves $\sigma^2(T)$ or $C(T)$ can thus be useful as a means of identifying the temperature range where careful cooling is required. They can actually be used while the algorithm proceeds (see § 3.4.2 and 3) to improve "dynamically" the parameters which control the annealing (L_t and temperature decrement).

3.5.3 *Average fluctuations of energy $|\Delta E(T)|$ for the accepted moves*

The use of this quantity, introduced by Darema *et al.* [6.3], is motivated by the fact that transitions are decided on the basis of energy differences between an old (before move) and a new (after move) configuration. Therefore small fluctuations in $<|\Delta E|>$ at high and intermediate temperatures may signal that the procedure got stuck in some high-energy local minimum. This quantity has been used to examine the behaviour of parallel SA algorithm.

3.5.4 *Acceptance ratio A(T)*

This parameter is actually an indicator of the efficiency of the exploration of the configuration space of the problem : it must be high at the beginning of the annealing and progressively drop to a small to negligeable value as temperature is decreased. It is of little use in the case of discrete problems except to signal when the system is getting frozen. For continuous variable problems, A(T) is also directly linked to the size of the elementary steps used to explore the configuration space. In Monte Carlo simulation of fluids it is usually considered that the optimal efficiency is obtained when approximately half of the transitions (moves) are rejected. It is possible to apply a similar criterion to SA [4.5]. More specifically an "acceptance window" is given and the step moves at each temperature are adjusted such that the acceptance rate remains between the limits of that window : if A exceeds the upper limit, the step size is increased; if it becomes smaller than the lower limit, the step size is decreased.

3.5.5 *Relaxation time ε(T)*

The relaxation time $\varepsilon(T)$ is a measure of the local time scale of the process and should ideally be an essential tool to estimate the optimum length of the Markov chains and thereby to devise an efficient annealing schedule [5.10,11,12]. Extensive numerical simulations for a TSP-type problem [5.10] have shown that $\varepsilon(T)$ is given to a good approximation by an Arrhenius type law. However evaluation of $\varepsilon(T)$ as the annealing proceeds is not feasible and this approach is not of much use in practice. To improve the efficiency of the algorithm.

3.6 Simulated annealing : do it yourself

The overall annealing procedure, as detailed in the previous sections can be summarized in the following pseudo-code :

```
begin
    Initialize
    t = 1
    do
        do
            Perturb the system :
            configuration ωᵢ --> configuration ωⱼ : ΔEᵢⱼ
            if ΔEᵢⱼ ≤ 0 then accept else
                if exp(-ΔEᵢⱼ/Tₜ) > random[0,1] then accept;
            if accept then Update (replace ωᵢ by ωⱼ)
        until equilibrium is approached closely enough
        Tₜ₊₁ = f(Tₜ)   (decrease temperature)
        t = t + 1
    until stop criterion is true
end
```

This illustrates the ease of implementation of the simulated annealing algorithm : it does not require the calculation of derivatives but only to formulate the cost value E as a function of the configuration of the system. However the cost function (or, more specifically, the change in cost ΔE_{ij}) is evaluated a large number of times as the algorithm proceeds and its calculation must be carefully optimized to save computing time.

4. PERFORMANCE OF THE ALGORITHM

For practical applications, a most prominent feature of SA is its ease of implementation. This is reflected in the wide variety of problems which have been handled with this method since 1983. However it is worthy to note that the performances of SA have been examined essentially in the case of combinatorial optimization problems while assessement of its performances in application to continuous problems are still very scarce. With respect to the results , the performance of the SA algorithm can be estimated by two quantities: the quality of the solution and the computing time required by the optimization.

4.1 Quality of the solution

The quality of the final solution which assesses the ability of the procedure to reach a near-globally optimum can be quantified by the difference in cost value between this solution and the true globally minimal configuration if it is known. Several theoretical analysis of the performance of the algorithm are available in the litterature (see e.g. Chapter 6 of ref. [2.1 & 2.2]). In addition extensive empirical performance analysis have been performed for a number of standard combinatorial optimization problems; although most of these studies have been performed with simple cooling schedules, they have shown that, with respect to the quality of the results, SA is quite competitive with and often outperforms tailored algorithm.

4.2 Running time

Conclusions concerning the running time of the algorithm are far more difficult to draw. Indeed, as mentioned earlier, a large fraction of the results available in the litterature have been obtained with conceptually simple (i.e. non optimized) cooling programs which certainly hides the merits of the algorithm. Even though the SA procedure, using computer time of the same order of magnitude, often performs better than tailored optimization algorithms.

To conclude this section, one must mention that the dependence of the final cost function on the cooling rate of the annealing procedure has been investigated by Aarts and Van Laarhoven in the case of the TSP and by Grest *et al.* [7.4], based on the analogy with model spin glasses. They have found that the lowest cost value obtained by SA depends logarithmically on the cooling rate Q : $E_{ground\ state} \approx -1/LnQ$.

5. APPLICATION TO OPTIMIZATION PROBLEMS

Since its introduction by Kirkpatrick *et al.* in 1983, the SA algorithm has been applied to a wide range of technical and scientific optimization problems. Surveying all the applications is outside the scope of this article (for a detailed annotated bibliography, see ref. 2.3). The following sections just try to provide a few references on the use of SA in a few areas of technological interest where SA has brought new results and/or significant improvements with respect to previous methods.

5.1 The Traveling Salesman Problem (TSP)

This is a classic example of a complex optimization problem (easy to state but hard to solve) and it has become a favorite playground for mathematicians. It can be stated as follows : "given a list of N cities with their location, find the shortest tour that visit each city exactly once". It is representative of a large class of optimization problems called NP-hard (or NP-complex) problems : such problems require a computational effort that grows faster than any power of their size (hence their name of Non-Polynomial). As such it has attracted a lot of effort to find polynomial-time algorithms giving approximate solutions. Most analysis of the performances of the SA algorithm are actually based on investigation of the TSP. Results obtained so far by SA compare well, in terms of both quality of the solution and computing time, with those obtained with tailored heuristics. For a working approach to SA applied to the TSP, see ref. 3.5.

Bibliography :
E. Bonomi & J.L. Lutton
The N-city traveling salesman problem : statistical mechanics and the Metropolis algorithm
SIAM Review **26**, 551-568 (1984)
R.E. Randelman & G.S. Grest
N-city traveling salesman problem : optimization by simulated annealing
J. Statistical Physics **45**, 885-890 (1986)
N. Sourlas
Statistical mechanics and the travelling salesman problem
Europhysics Letters **2**, 919-923 (1986)
E.H.L. Aarts, J.H.M. Korst & P.J.M. van Laarhoven
A quantitative analysis of the simulated annealing algorithm : a case study for the traveling
salesman problem
J. Statistical Phys. **50**, 187-206 (1988)

5.2 VLSI placement and routing problems

The development of modern electronic systems is based on the use on high-density very large scale integrated (VLSI) circuits. The design of such circuits is a very lengthy and laborious task and a number of techniques have been developed to automate various steps of the process. Very schematically, the design optimization can be divided in two steps : a layout problem (placing modules of variable size in an area of minimal size) and a routing one (connect the modules by wires of minimal length). Application of SA in this field has been extremely successfull : a large fraction of the publications on the algorithm and the majority of its applications actually belong to the area of computer-aided circuit design. A few early papers on the subject are quoted below; for a detailed bibliography on the matter consult reference [2.3].

Bibliography :
M.P. Vecchi & S. Kirkpatrick
Global wiring by simulated annealing
IEEE Transactions on Computer-Aided Design **2**, 215-222 (1983)
P. Siarry & G. Dreyfus
An application of physical methods to the computer aided design of electronic circuits
J. Physique Lett. **45**, L39-L48 (1984)
S.A. Kravitz & R. Rutenbar
Placement by simulated annealing on a multiprocessor
IEEE Transactions on Computer-Aided Design **6**, 534-549 (1987)

5.3 Image processing

The restoration of degraded images is an optimization problem of evident practical importance; Geman & Geman were the first to propound a generalization of SA to find a maximum a-posteriori distribution for an image corrupted by noise. Their approach has prompted considerable interest in this area. The main shortcoming of the method is its slow convergence but it can be possible to overcome by a parallel implementation of the algorithm.

Bibliography :

S. Geman & D. Geman
Stochastic relaxation, Gibbs distributions and the Bayesian restoration of images
IEEE Trans. PAMI **5**, 721-741 (1984)
P. Carnevalli, L. Coletti & S. Patarnello
Image processing by simulated annealing

IBM J. Res. Develop. **29**, 569-579 (1985)
G. Wolberg & T. Pavlidis
 Restoration of binary images using stochastic relaxation with annealing
Pattern Recognition Letters **3**, 375-388 (1985)

6. APPLICATIONS TO PROBLEMS OF STRUCTURE DETERMINATION

Application of SA to crystallographic problems has been rather limited so far. It was initiated, independantly from Kirkpatrick and Cerny's work, by russian scientists but remained apparently limited to a single feasibility test. More recently, SA has been applied in protein crystallography for both search and refinement problems.

6.1 Crystal structure determination from diffraction data

The use of simulated annealing (the name had not been invented yet !) to solve the phase problem has been proposed as early as 1979 by Khachaturyan *et al.* but, strangely enough, this work has remained widely overlooked. The determination of crystal structures from diffraction data reduces to finding the atomic scale arrangement within a triply periodic unit-cell that gives the best agreement between observed and calculated diffraction intensities (structure factors). This agreement is usually quantified by the R-factor :

$$R = \sum_{hkl} | | F_{obs.}(hkl)| - |F_{calc.}(hkl)| | / \sum_{hkl} |F_{obs.}(hkl)|$$

or any related parameter which tends to zero for the arrangement corresponding to the correct structure. In optimization terms, the problem of structure determination can thus be re-phrased as that of locating the global minimum of R in the multidimensional space of the independent atomic coordinates; owing to the numerous local minima occuring in this space, local optimization methods (e.g. least-squares) are clearly inefficient for structure determination and apply only to structure refinement. Using the R-factor as a cost function (and a crude annealing algorithm) Khachaturyan *et al.* have tested the method on a known structure containing 8 independent atoms (24 variables) in an orthorhombic unit cell. Although they had to discretize the problem to limit the computing time (trial atomic positions are located on a grid of about 0.3Å spacing), their success at reaching a topologically correct solution shows that the global approach to optimization given by SA provides a new method of automatic crystal structure determination. A non negligeable interest of this approach to structure determination and refinement (in principle the algorithm does both at the same time although a conventional least-squares refinement would be more efficient in the final stages of the procedure) is that it enables to include easily all a-priori information which are available to constrain the optimization : this only requires to express the constraints as a function of the atomic positions and to include them (with a proper weighting) into the total cost function.

Bibliography :
A.G. Khachaturyan, S.V. Semenovskaya & B. Vainshtein
Statistical-thermodynamic approach to determination of structure amplitude phases
Kristallografiya **24**, 905-916 (1979); *Sov. Phys. Crystallogr.* **24**, 519-524 (1979)
A.G. Khachaturyan, S.V. Semenovskaya & B. Vainshtein
The thermodynamic approach to the structure analysis of crystals
Acta Cryst. A **37**, 742-754 (1981)
S.V. Semenovskaya, K.A. Khachaturyan & A.G. Khachaturyan
Statistical mechanics approach to the structure determination of a crystal
Acta Cryst. A **41**, 268-273 (1985)

6.2 Crystallographic refinement of protein structures

Refinement of macromolecular crystal structures from diffraction data is still a laborious task which usually consists of two steps :

- a conventional least-squares refinement (often including some kind of stereochemical and internal packing restraints) which minimizes the discrepancy between observed and calculated structure factors
- a "manual" model rebuilding using interactive computer graphics. This stage is required to correct the positions of residues that are misplaced (local traps) by a distance larger than the radius of convergence of the least-squares refinement (typically 1 Å).

This two-step process (refered to as l.s. refinement + model building) must be repeated a number of times which requires a lot of computing time and human effort. In an attempt to keep manual intervention to a minimum, Brünger *et al.* have proposed to introduce SA into the refinement procedure. The cost function to optimize is defined as the weighted sum of an empirical potential energy term, describing stereochemical and non-bonded interactions, and the R-factor. However the approach followed by Brünger *et al.* differs significantly in many respects from the usual simulated annealing one : first, in contrast with all previous uses of SA, these authors have suggested to explore the configuration space by using molecular dynamics instead of the usual Metropolis importance sampling. This choice is justified by the argument that " for large biomolecular structures, the molecular dynamics algorithm is generally more efficient at generating equilibrium structures". This choice probably deserves further discussion but, as far as the SA algorithm is concerned, it obviously raises a difficulty : temperature is not constant during the simulation ! This technical problem is overcome by periodically rescaling the velocity of all atoms in the system to keep temperature constant. A second difference between this approach and standard SA is that the number of "temperature steps" is limited to two or three only. The question of the relationship between this procedure (probably better termed crystallographic refinement using molecular dynamics) and simulated annealing is thereby still open (for a discussion of the combined use of molecular dynamics and simulated annealing, see for instance R. Car & M. Parrinello, *Phys. Rev. Lett.* **55**, 2471 (1985)). Nevertheless the method is claimed to "save a great deal of human effort, albeit at a non negligible cost of super computer time" (Kuriyan *et al.*).

Using a more traditional SA approach Subbiah and Harrison have addressed a search problem which is a crystallographic analog of the placement problem mentioned above. The goal of this search problem (often encountered in protein crystallography) is to "place a rigid molecular structure in an electron density map in the most objective fashion". With the advent of powerful computers this is usually performed by exhaustive multi-dimensional searches. The application of SA to this problem has been extremely successful on two respects :
- in cases where both the exhaustive search and the SA procedure could be used, the latter produced a major proportion of all good solutions obtained by the former
- the computing time is drastically reduced : from about 6.5 hours of CPU time on a Micro-Vax II to 15 minutes for the placement of a sugar in the bonding-site electron density. As the improvement increases with the size of the problem, the use of the SA procedure allows to conduct searches that were impossible by exhaustive search with current computers.

Bibliography :
A.T. Brünger, J. Kuriyan & M. Karplus
Crystallographic R factor refinement by molecular dynamics
Science **235**, 458-460 (1987)
A.T. Brünger
Crystallographic refinement by simulated annealing
in *Crystallographic Computing 4, Techniques and new technologies*, p.126-140, N.W. Isaacs & M.R. Taylor Ed., International Union of Crystallography, Oxford University Press (1988)
A.T. Brünger
Crystallographic refinement by simulated annealing : application to a 2.8 resolution structure of aspartate aminotransferase
J. Mol. Biol. **203**, 803-816 (1988)
A.T. Brünger, M. Karplus & G.A. Petsko
Crystallographic refinement by simulated annealing : application to cramoin
Acta Cryst. A **45**, 50-61 (1989)
S. Subbiah & S.C. Harrison

A simulated annealing approach to the search problem of protein crystallography
Acta Cryst. A **45**, 337-342 (1989)
J. Kuriyan, A.T. Brünger, M. Karplus & W.A. Hendrickson
X-ray refinement of protein structures by simulated annealing : test of the method on Myohemerythrin
Acta Cryst. A **45**, 396-409 (1989)
M. Fujinaga, P. Gros & W.F. van Gunsteren
Testing the method of crystallographic refinement using molecular dynamics
J. Appl. Cryst. **22**, 1-8 (1989)

6.3 Molecular conformation of molecular clusters

Molecular conformation of atomic clusters held together by two-body forces is usually obtained theoretically by energy minimization of their geometry using simple interatomic potentials (e.g. Lennard-Jones). For large clusters, this approach suffers from the fact that the number of low-lying energy minima grows almost exponentially with the size of the problem. This is therefore a typical problem for which SA can bring new results at a minimum computing cost investment. Wille has undertaken a thorough search of the stable conformations for Lennard-Jones clusters in the size range $4 \leq N \leq 25$. In all cases, SA reproduced the previously known geometries; in one case (N=24), it produced a structure with an energy lower than that previously obtained by other methods. The procedure can easily been extended to minimum energy search in higher dimensions.
A related problem of geometry optimization is discussed by Donnelly.

Bibliography :
L.T. Wille & J. Vennik
Computational complexity of the ground-state determination of atomic clusters
J. Phys. **A18**, L1419-L1422 (1985)
L.T. Wille
Minimum-energy configurations of atomic clusters : new results obtained by simulated annealing
Chem. Phys. Letters **133**, 405-410 (1987)
L.T. Wille
Close packing in curved space by simulated annealing
J. Phys. **A20**, L1211-L1218 (1987)
R.A. Donnelli
Geometry optimization by simulated annealing
Chem. Phys. Letters **136**, 274-278 (1987)

6.4 Equilibrium configuration of a set of charges

The problem of the equilibrium of N equal point charges in coulombic interaction within a circle or a sphere (or any other polyhedral box) sounds like an easy and somewhat old-fashioned question. Indeed it has been solved for small values of N (<12). However from a computational point of view, the minimization problem becomes very complex for large values of N and was solved only recently, for values up to 50, by Wille and Vennik using a SA algorithm. In addition to proving the interest of SA to solve continuous problems of that kind, their work also emphasizes one of the difficulties of the method : first calculations led to several "equilibrium configurations" which later turned out to correspond to local minima rather than true ground state; the correct ground states were obtained by the same method using more careful annealing. This example also underscores a fundamental feature of optimization, namely the notion that "there is no general way to prove a configuration is a global minimum" [3].

Bibliography :
L.T. Wille & J. Vennik
Electrostatic energy minimization by simulated annealing

J. Phys. A : Math. Gen **18**, L1113-L1117 (1985), Corrigendum ibid. **19**, 1983 (1986)

6.5 Determination of magnetic structures

Monte Carlo is a widely used technique to investigate the behavior of magnetic systems, especially phase transitions and critical phenomena (see for instance [2]). In order to obtain statistically significant results, such applications require to use large simulation samples and are rather computer time demanding. Used within a SA algorithm and with limited size samples, Monte Carlo also enables to determine the ground state configuration of magnetic models and, in particular, to predict magnetic structures from postulated coupling constants. It is important to stress that, for such applications and in contrast with usual Monte Carlo studies, one does not investigate the evolution of the sample under the influence of temperature (e.g. to locate phase transitions) but rather aims at finding the spin configuration which agrees best with a set of constraints (the hamiltonian of the spin system) <u>and</u> a given unit-cell. This approach, developed by Lacorre [4], has proved to be very effective to solve the structure of several highly frustrated insulating compounds for which coupling constants were fairly accurately known. To apply the SA algorithm, one must provide the following elements :
- a description of the topology of the magnetic lattice, i.e. give all spins of the simulation box with a list of their neighbours together with the corresponding coupling constants
- the hamiltonian of the system which, in general, is the sum of three terms :

$$E = - \sum_{<i,j>} {}^t S_i \, [J]_{ij} \, S_j + \sum_i \{ D_i \, (r_i.S_i / |r_i|)^2 - {}^t H.S_i \}$$

representing respectively a two-spin coupling term with (possibly) an anisotropic asymmetric exchange tensor, a single-ion anisotropy term and an applied magnetic field term - an annealing schedule; the simple cooling law used [4] is controled by a single parameter, the cooling rate α. The constant length of the Markov chains is determined by trial-and-error.
In addition to its use at determining magnetic structures, the program which accepts any kind of 1D, 2D or 3D topology provides a very flexible technique to investigate the ground state configuration of magnetic models (e.g. magnetic phase diagrams).

Bibliography :
P. Lacorre & J. Pannetier
MCMAG : a program to simulate magnetic structures
J. Magn. Magn. Mat. **71**, 63-82 (1987)
P. Lacorre
MCMAG Manual (Version 88.01) ILL Technical Report 88LA13T (1988)
[The Fortran program and manual are available from the author upon request]

6.6 Prediction of crystal structures from crystal-chemistry rules

The usual approach to crystal structure modelling is to minimize, for a selection of candidate crystal architectures, the potential energy of the system with respect to the structural parameters of these models : the solution is the arrangement which comes out lower in energy. Such procedures, which are normally based on the use of empirical or ab-initio pair potentials, are normally restricted to local *structure optimization* within the constraints of given symmetry and bond topology [see e.g. ref. 5] and do not really address the problem of predicting the unknown structure of a real compound. For practical application, the latter problem can usefully be stated in the following way : given the chemical composition and unit-cell of a crystalline material, find all structural arrangements of atoms which fulfill a set of geometrical and/or crystal chemistry rules. In the SA approach to this question, the reasonableness of a given structural arrangement within the experimentally known unit-cell is described in terms of a cost function which quantifies its departure from the rules.
A first method, based on geometrical arguments, has been proposed recently by Deem and Newsam for crystal structures built up from a framework of vertex-sharing tetrahedra; their approach is based on the observation that, for all known zeolites structures, the distances T-T

and the angles T-T-T, where T represents the tetrahedral species (Si/Al in zeolites), vary over rather limited ranges. Therefore a cost function can be derived empirically from histogrammed data of T-T distances and T-T-T angles taken from a selection of representative zeolite structures; the configuration space to explore is reduced by making use of symmetry elements. The coordinates of the T atoms are determined by minimizing the cost function using SA; the position of the framework oxygen atoms can then be obtained by conventional modelling such as Distance-Least-Squares (DLS) optimization [6]. The method has been found to be very effective for problems of small size (up to about 6 independent T atoms per unit-cell); for larger size problems, the crudeness of the cost function apparently gives rise to a large number of almost equivalent solutions (degeneracy) which makes the method less efficient.

In the case of ionic materials, a more effective cost function can be derived from Pauling's second rule [7] written in its bond valence/bond length formulation [8] :

$$E_{valence} = \sum_i^{cell} |q_i - \sum_{j=1}^{CN_i} s_{ij} |$$

where q_i is the valence (formal charge) of ion i, CN_i its coordination number and s_{ij} the valence (strength) of bond ij; s_{ij} is commonly linked to interatomic separation r_{ij} by empirical expressions of the form $s_{ij} = (r_{ij}/r_0)^{-N}$ or $s_{ij} = \exp[(r_0 - r_{ij})/B]$ where r_0, N and B are constants for a particular ion pair. In contrast with most energy functions used in structure modelling, the above expression does not refer to single pair interactions but involve the complete coordination sphere of every atom in the structure. This rule of local equilibrium which controls the coordination of cations by anions and vice-versa does not however restrict either cation-cation or anion-anion packing. An elementary although not general method to constrain these two packings is through an electrostatic (repulsive) interaction of the form :

$$E_{electr.} = \frac{1}{2} \sum_{i,i'}^{cell} \frac{q_i q_{i'}}{r_{ii'}}$$

where the indices i,i' run on all cation-cation and anion-anion pairs. Introducing a coefficient λ to express the relative cost of the two contributions, the total cost function writes :

$$E = (1-\lambda).E_{valence} + \lambda.E_{electr.}$$

The choice of λ turned out to be rather critical but that it can be conveniently estimated during the early stages of the simulation from the variance σ^2 of the two terms of the cost function :

$$\lambda = \sigma^2_{valence} / [\sigma^2_{valence} + \sigma^2_{electr.}] \quad where \quad \sigma^2_k = <E_k^2> - <E_k>^2.$$

Structure determination then reduces in locating the configuration of ions which yields the global minimum of the cost function E. The potential of this algorithm has been tested on a series of simple structures and found to produce a correct solution with a high probability. The main limitation of the current program results from the inaccuracy of the bond valence parameters currently available as well as from the inadequacy of the second term of the cost function to represent a large fraction of known structures (e.g. topologies exhibiting face-sharing polyhedra).

In spite of their limitations, these two exemples demonstrate that SA is a powerful tool to identify structural models compatible with a given set of building principles. As such this approach can provide a useful and flexible modelling method : the challenge that remains for the moment is to conjecture the empirical or semi-empirical rules which govern the architecture of the solids.

Bibliography :
M.W. Deem & J.M. Newsam
Determination of 4-connected framework crystal structures by simulated annealing
Nature **342**, 260-262 (1989)
J. Pannetier, J. Bassas-Alsina, J. Rodriguez-Carvajal & V. Caignaert
Prediction of crystal structures from crystal chemistry rules by simulated annealing
Submitted for publication

7. CONCLUSIONS

In this article, we have tried to give an introduction to the use and applications of the simulated annealing algorithm. The proliferation of publications mentionning SA is an indication of its ease of implementation and apparent universality; the essential drawback remains its "gargantuan appetite for cpu time". Improvements of the efficiency of the procedure in the future can be foreseen in the following aspects :
- improvement of the generation mechanism (concept of move classes [5.1], rejectionless method [5.6], Langevin molecular dynamics [8.5]) in order to explore more efficiently the configuration space of the system being optimized, thus speeding up execution
- parallel implementation of the algorithm : although the nature of the SA algorithm greatly hampers the design of parallel algorithms, an approach based on neural networks (Boltzmann machines) may open the way to new developments.

List of references

1 -   N. Metropolis, A. Rosenbluth, M. Rosenbluth, A. Teller & E. Teller
      J. Chem. Phys. **21**, 1087-1092 (1953)
2 -   K. Binder
      Topics in current Physics, Vol. **36**, Springer Verlag Berlin, (1984)
3 -   M.J. Sabochick & D.C. Richlin
      Phys. Rev. **37**, 10846-10850 (1988)
4 -   P. Lacorre, Thesis, Université du Maine (1988)
5 -   W.R. Busing, WMIN, a computer program to model molecules and crystals in terms of
      potential energy functions, ORNL - 5497 (1981)
6 -   W.M. Meier & H. Villiger, Z. *Kristallogr.* **129**, 411-423 (1969)
7 -   L. Pauling, *J. Amer. Chem. Soc.* **51**, 1010-1026 (1929)
8 -   I.D. Brown, in *Structure and Bonding* (eds. M. O'Keeffe & A. Navrotsky) **2**, 1-30
      (Academic Press, New-York 1981)

Acknowledgements

Thanks are due to A. Antoniadis, P. Lacorre and J. Rodriguez-Carvajal for many stimulating discussions on the theory and applications of simulated annealing.

A selected list of references on Simulated Annealing

Although it all started in 1983, more than 300 papers and reports have now appeared on both the theory and the applications of SA; owing to the widespread interest for the method, these papers are scattered in nearly a hundred journals. The following list does not intend to be comprehensive but rather to provide the minimum information required to get started (1, 2 & 3) together with a set of more specialized references (4 to 8).

1 -   The origins

1   S. Kirkpatrick, C.D. Gelatt & M.P. Vecchi
     Optimization by simulated annealing
     Science **220**, 671-680 (1983)
2   V. Cerny
     Thermodynamical approach to the traveling salesman problem : an efficient simulation algorithm
     J. Optimization Theory Applic. **45**, 41-51 (1985)

These are the two seminal (and very readable) papers which introduced the very concept of SA for optimization problems. A "must" to get started !

2 -   Reviews : several review papers on SA have been published by Aarts and Van Laarhoven. Their book :

1   P.J.M. van Laarhoven & E.H.L. Aarts
     Simulated Annealing : Theory and Applications
     Kluwer, Dordrech (1987)

gives a state of the art overview of the theory and applications of SA; it is recommended to everyone who really wishes to apply SA to specific problems. A shorter presentation is given in the first part of :

2   E.Aarts & J. Korst
     Simulated annealing and Boltzmann machines
     Wiley - Interscience Series in Discrete Mathematics and Optimization (1989)

which also discusses the link between SA and neural computing. Recent applications of SA to engineering optimization problems are reviewed in :

3   *Simulated Annealing (SA) and Optimization : modern algorithms with VLSI, optimal design, & missile defense applications*
     Edited by M.E. Johnson, American Sciences Press, New-York (1988)

This book, which is a special issue of the American Journal of Mathematical and Management Sciences (Vol. 8, Nos. 3 & 4, 1988), also includes a detailed annotated bibliography on simulated annealing.

3 -   Introductory papers

1   S.Kirkpatrick
     Optimization by simulated annealing : quantitative studies
     J. Statistical Phys. **34**, 975-986 (1984)
2   B. Hajek
     A tutorial survey of theory and applications of simulated annealing
     Proc. of the 24th Conference on Decision & Control, Ft. Lauderdale (Dec. 1985), p. 755-760

3   I. Morgenstern & D. Würtz
     Simulated annealing for "spin-glass-like" optimization problems
     Z. Phys. B - Condensed Matter **67**, 397-403 (1987)
4   E.H.L. Aarts & P.J.M. van Laarhoven
     A pedestrian review of the theory and application of the simulated annealing algorithm
     in the proceedings of "Spin glasses, optimization and neural networks", Ed. J.L. van
     Hemmen & I. Morgenstern - Lecture Notes in Physics 275 - Springer Verlag 1987

     A brief introduction and a working approach (applied to the traveling salesman problem)
     are given in section 10.9 of the very helpful :
5   W.H. Press, B.P. Flannery, S.A. Teukolsky & W.T. Vetterling
     Numerical Recipes
     Cambridge University Press, 1989

4 -   Applications of SA to continuous problems

The theory and applications of SA have been developed essentially in the case of discrete
problems. Application to optimization problems with continuous variables is far less
documented and mastered. A few references are given below :

1   D. Vanderbilt & S.G. Louie
     A Monte Carlo simulated annealing approach to optimization over continuous variables
     J. Comput. Phys. **56**, 259-271 (1984)
2   A. Khachaturyan
     Statistical approach in minimizing a multivariate function
     J. Math. Phys. **27**, 1834-1838 (1986)
3   I.O. Bohachevsky, M.E. Johnson & M.L. Stein
     Generalized simulated annealing for function optimization
     Technometrics **28**, 209-217 (1986)
4   S. Geman & C.R. Hwang
     Diffusions for global optimization
     SIAM Journ. of Control and Optimization **24**, 1031-1043 (1986)
5   A. Corana, M. Marchesi, C. Martini & S. Ridella
     Minimizing multimodal functions of continuous variables with the "simulated annealing"
     algorithm
     ACM Trans. Math. Software **13**, 262-280 (1987)
6   H. Szu & R. Hartley
     Fast simulated annealing
     Physics Letters A **122**, 157-162 (1987)
7   D.G. Brooks & W.A. Verdini
     Computational experience with generalized simulated nnealing over continuous variables
     Am. J. Mathematical Management Sciences **8**, 425-449 (1988)

5 -   Mathematical analysis of SA (optimization of the algorithm)

The determination of an adequate cooling schedule is an essential problem for practical
applications since it determines both the quality of the final solution and the computing
time needed to reach it. It has been addressed in many papers over the last five or six
years. A few significant approaches are discussed in the following papers :

1   S.R. White
     Concepts of scale in simulated annealing
     Proceedings IEEE Int. Conf. on Computer Design, Port Chester (Nov. 1984), 646-651
2   S. Geman & D. Geman
     Stochastic relaxation, Gibbs distributions and the Bayesian restoration of images
     IEEE Trans. PAMI **5**, 721-741 (1984)

3   F. Romeo, A.L. Sangiovanni-Vincentelli & C. Sechen
       Research on simulated annealing at Berkeley
       Proceedings IEEE Int. Conf. on Computer Design, Port Chester (Nov. 1984), 652-657
4   B. Gidas
       Nonstationary Markov chains and convergence of the annealing algorithm
       J. Statistical Phys. **39**, 73-131 (1985)
5   D. Mitra, F. Romeo & A. Sangiovanni-Vincentalli
       Convergence and finite-time behavior of simulated annealing
       Adv. Appl. Prob. **18**, 747-771 (1986)
6   J.W. Greene & K.J. Supovit
       Simulated annealing without rejected moves
       IEEE Trans. on Computer-Aided Design **5**, 221-228 (1986)
7   M.D. Huang, F. Romeo & A. Sangiovanni-Vincentalli
       An efficient general cooling schedule for simulated annealing
       Proceedings IEEE Int. Conf. on Computer-Aided Design, Santa Clara (Nov. 1986),381-384
8   S. Anily & A. Federgruen
       Simulated annealing methods with general acceptance probabilities
       J. Appl. Prob. **24**, 657-667 (1987)
9   S. Anily & A. Federgruen
       Ergodicity in parametric nonstationary Markov chains : an application to simulated annealing methods
       Operat. Res. **35**, 867-874 (1987)
10   S. Rees & R.C. Ball
       Criteria for an optimum simulated annealing schedule for problems of the travelling salesman type
       J. Phys. A: Math. Gen. **20**, 1239-1249 (1987)
11   J.D. Nulton & P. Salamon
       Statistical mechanics of combinatorial optimization
       Phys. Rev. A **37**, 1351-1356 (1988)
12   B. Andresen, K.H. Hoffmann, K. Mosegaard, J. Nulton, J.M. Pedersen & P. Salamon
       On lumped models for thermodynamic properties of simulated annealing problems
       J. Phys. France **49**, 1485-1492 (1988)
13   R. Holley & D. Stroock
       Simulated annealing via Sobolev inequalities
       Commun. Math. Phys. **115**, 553-569 (1988)
14   B. Hajek
       Cooling schedules for optimal annealing
       Math. Operat. Res. **13**, 311-329 (1988)

6 -   Parallel algorithms and SA

The principal shortcoming of SA is the amount of computational effort it requires to converge to a near-optimal solution. A possible method to increase the speed of the algorithm is to distribute the execution of its different steps over several interconnected parallel processors. The design and performances of parallel SA algorithms are discussed in the following articles :

1   D.W. Murray, A. Kashko & H. Buxton
       A parallel approach to the picture restoration algorithm of Geman and Geman on an SIMD machine
       Image and Vision Computing **4**, 133-142 (1986)
2   E.H.L. Aarts, F.M.J. De Bont, J.H.A. Habers & P.J.M. Van Laarhoven
       Parallel implementation of the statistical cooling algorithm
       Integration **4**, 209-238 (1986)

3   F. Darema, S. Kirkpatrick & V.A. Norton
     Parallel algorithms for chip placement by simulated annealing
     IBM J. Res. Develop. **31**, 391-402 (1987)
4   A. Casotto, F. Romeo & A.L. Sangiovanni-Vincentelli
     A parallel simulated annealing algorithm for the placement of macro-cells
     IEEE Trans. on Computer-Aided Design **6**, 838-847 (1987)

For a general discussion of this aspect of SA, consult Chapter 6 of ref. [2.2]

7 - Relation between SA and statistical physics

As mentioned in § 2, the use of SA to solve large combinatorial optimization problems is based on an analogy with the annealing of solids. Following the successes of SA in optimization, a number of authors have elaborated on the relation between statistical mechanics and combinatorial optimization; this is of interest both from a fondamental viewpoint and to improve the efficiency of the algorithm for practical applications.

1   J. Vannimenus & M. Mezard
     On the statistical mechanics of optimization problems of the traveling salesman type
     J. Physique Lett. **45**, L-1145-1153 (1984)
2   S.Kirkpatrick & G. Toulouse
     Configuration space analysis of travelling salesman problems
     J. Physique **46**, 1277-1292 (1985)
3   R. Ettelaie & M.A. Moore
     Residual entropy and simulated annealing
     J. Physique Lett. **46**, L893-900 (1985)
4   G.S. Grest, C.M. Soukoulis & K. Levin
     Cooling-rate dependance for the spin-glass ground state energy : implications for optimization by simulated annealing
     Phys. Rev. Lett. **56**, 1148-1151 (1986)
5   A. Khachaturyan
     Statistical approach in minimizing a multivariate function
     J. Math. Phys. **27**, 1834-1838 (1986)
6   Yaotian Fu & P.W. Anderson
     Application of statistical mechanics to NP-complete problems in combinatorial optimization
     J. Phys. A: Math. Gen. **19**, 1605-1620 (1986)
7   M. Mézard
     Spin glasses and optimization
     in *Lecture Notes in Physics* **275**, J.L. van Hemmen & I. Morgenstern eds., 354-372 (1987)
     Springer-Verlag (Berlin)
8   I. Morgenstern
     Spin glasses, optimization and neural networks
     in *Lecture Notes in Physics* **275**, J.L. van Hemmen & I. Morgenstern eds., 399-427 (1987)
     Springer-Verlag (Berlin)

8 - Further applications in condensed matter physics

Application of SA algorithm in physics still are few. In addition to the problems presented in section 6 of this review, a few references are mentioned below :

1   D.M. Nicholson, A. Chowdary & L. Schwartz
     Monte Carlo optimization of pair distribution functions : application to the electronic structure of disordered metals

Phys. Rev. B 19, 1633-1637 (1984)

2   D.H. Rothman
Nonlinear inversion, statistical mechanics and residual statics estimation
Geophysics 50, 2784-2796 (1985)

3   F. Wooten, K. Winer & D. Weaire
Computer generation of structural models of amorphous Si and Ge
Phys. Rev. Lett. 54, 1392-1395 (1985)

4   A. Lyberatos, P. Wohlfarth & R.W. Chantrell
Simulated annealing : an application in fine particles magnetism
IEEE Trans. on Magnetics 21, 1277-1282(1985)

5   R. Biswas & D.R. Hamann
Simulated annealing of silicon clusters in Langevin molecular dynamics
Phys. Rev. B 34, 895-901 (1986)

6   H. Telley, T.M. Liebling & A. Mocellin
Reconstruction of polycrystalline structures : a new application of combinatorial optimization
Computing 38, 1-11 (1987)

7   I. Heynderickx & H. De Raedt
Calculation ofthe director configuration of nematic liquid crystals by the simulated annealing method
Phys. Rev. A 37, 1725-1730 (1988)

Inst. Phys. Conf. Ser. No 107: Chapter 1
Paper presented at Neutron Scatt. Data Anal. Conference, Rutherford Appleton, 1990

45

The Bayesian approach to optimal instrument design

D. S. Sivia, R. N. Silver and R. Pynn
Theoretical Division and Manuel Lujan Jr. Neutron Scattering Center
Los Alamos National Laboratory
Los Alamos, NM 87545, U.S.A.

ABSTRACT: In data analysis we are faced with the task of making inferences about some quantity of interest, given experimental measurements. The use of entropic and other analysis methods is a case of trying to do the best with the data we have. A complementary, and perhaps even more important, issue is the question of how to design instrumentation to obtain "better" data. We present an introductory tutorial on the Bayesian approach to this type of question. Specifically, we show how a simple Bayesian analysis leads to the derivation of a much more robust figure-of-merit for the optimisation of an instrumental resolution function than is commonly used, with potentially serious implications for the design of neutron scattering facilities.

1. Introduction

In neutron scattering, we often wish to learn about the scattering law for a sample given measurements of the neutron counts in the various detectors of a diffractometer. The use of maximum entropy (MaxEnt), and other techniques, in the analysis of the data is a case of trying to do the best with the data we have. Usually, this is all that can be done: a user goes to do an experiment at a facility like LANSCE; since the instrumentation and hardware already exists, often the only freedom he or she has in terms of the data collected is the time allocated to do the experiment (which governs the statistical accuracy of the data). Let us suppose, however, that we are going to build a new facility, or just a new spectrometer. How should we design it to get the "best" data? This is an important question since a new facility can cost a hundred million dollars or more (even a single spectrometer can cost a million or two)!

Silver, Sivia & Pynn (1989) have addressed this question from a heuristic view-point, and have also suggested a quantitative answer based on elementary signal-to-noise ratio arguments from a power spectrum error analysis (Sivia, Silver & Pynn 1990). They posed the following question: "Given that the neutron scattering data are usually a blurred and noisy version of the scattering law we want, what are the optimal characteristics of the instrumental resolution function?" Conventional wisdom suggests that the most important characteristic of the resolution function is its width: the wider the resolution function, the poorer the quality of the data in the sense that it is more difficult to determine reliably the underlying scattering law. This thinking is based on a visual, or "what-you-see-is-what-you-get", consideration of the data. A more formal statistical inference, or image processing, analysis leads us to the conclusion that the overall shape of the resolution function is more important than its width.

In this paper we outline the Bayesian approach to this question of optimal instrument design, the algebra being found in Sivia (1990). We will formulate the problem in the same way as did Silver et al. and, as such, merely provide a Bayesian rationale for their results. The real advantage of the Bayesian approach is that it is far more general in its use, an almost identical analysis being applicable for the question of experimental design in many other contexts.

Although we will consider the design question for the specific case of neutron scattering, the conclusions will be valid for instrumental setups in many different scientific fields since most experiments involve at least some element of an instrumental resolution.

In Section 2 we briefly review the traditional procedure for designing an instrument and in Section 3 we outline the Bayesian approach for its optimisation. In Section 4 we consider three specific examples for optimising the instrumental resolution function: (i) the simplest case, which leads to the same optimisation criterion as advocated by conventional thinking; (ii) the second most simple case, which highlights a basic difficulty in what we mean by optimisation; (iii) a generalised case, which leads to the derivation of more robust figure-of-merit than is commonly used. We conclude with Section 5.

2. Designing an instrument

How do people design instruments? Our friend Jack Carpenter told us: "Well, they write large monte carlo programs to simulate their proposed instruments.They take 'typical' spectra (or scattering laws) and see how the instrumental parameters like flight-path lengths, collimation angles, and so on, affect the resultant data; they vary the instrumental parameters in order to 'improve' the quality of the data. Six months later their boss confronts them asking why they don't have a hole-in-the-ground to show for their work, and tells them to go and build something!" This procedure is schematically sketched in Fig. 1.

But, what do we mean by "improving the data"? This is the crux of the problem. It seems to us that traditionally the optimisation has been carried out visually — just look at the resulting data and see which one appears to be better. This visual optimisation leads to: (i) a preference for symmetric-looking data; (ii) a conventional figure-of-merit based on the total number of neutrons measured and the full-width-half-maximum of the resultant instrumental resolution function. Consider the example of Fig. 2: we show three sets of data generated by convolving the same "typical" spectrum with three different resolution functions (corresponding to three different choices of instrumental parameters). Which is the best? Conventional wisdom would probably say that 2(a) is much better than either 2(b) or 2(c) because the resolution function for 2(a) is much narrower than for 2(b) & 2(c). The total number of neutrons in 2(b) & 2(c) are comparable, and so are the widths of their resolution functions; since the data in 2(c) look lop-sided, we would probably prefer 2(b) to 2(c). We will return to the merit, or otherwise, of the this traditional optimisation strategy later, but let us now introduce the Bayesian approach to this question. (For a general background to Bayesian ideas see, for example, Jeffreys 1939 or Jaynes 1986.)

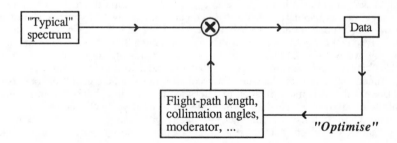

Figure 1: A diagram representing the way instruments are designed. A monte carlo program is written to simulate the proposed instrument in software; the instrumental parameters are varied to see their affect on data that would result from a "typical" spectrum (or scattering law). The crux of the problem is how to "optimise" the design?

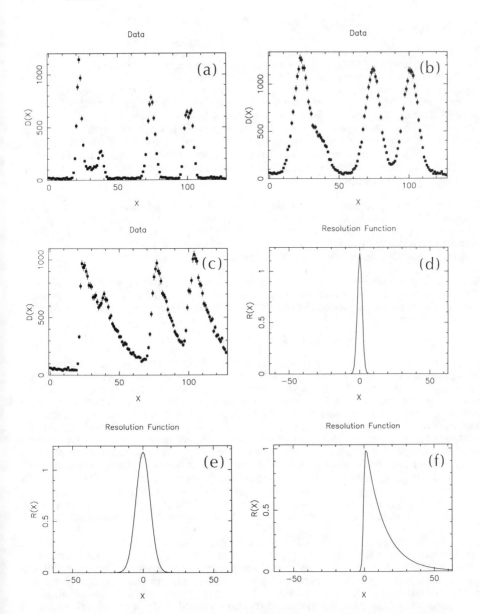

Figure 2: **(a)-(c)** are three sets of data generated by convolving the same "typical" spectrum with three different resolution function, **(d)-(f)** respectively, representing three choices of instrumental parameters. Which is best?

3. The Bayesian approach

We said at the outset that in data analysis we are faced with the task of making inferences about some quantity of interest, given experimental measurements. We can summarise this inference with a conditional probability distribution; for the case of a neutron scattering experiment, for example, this would be *prob[Scattering law / Data, Experimental setup]*, where "/" means "given". The larger the probability assigned to a particular scattering law, the more likely we believe it is to be the true scattering law in the light of the data. Our best estimate of the scattering law is given by that scattering law for which this probability distribution is a maximum, and the reliability of this estimate is given by the width or spread of the probability distribution about its maximum. If this probability distribution is sharply-peaked, we are very confident of our prediction; if it is very broad, we are fairly uncertain about the true scattering law.

The criterion for optimising instrumental design can thus be addressed by the following question: "Given that the data we measure are dependent on the experimental setup, how should we design our instrument (spectrometer & moderator) to obtain the most *reliable* estimate of the scattering law?" The answer is rather obvious: we need to make the probability distribution prob[Scattering law I Data, Experimental setup] sharply-peaked!

But, how does the experimental setup affect the width of this probability distribution? Well, the answer is given by Bayes' theorem. Bayes' theorem tells us how the probability distribution we require is related to two others, one of which can be calculated and the other "guessed". It states that:

$$\text{prob}[S|D,E] \propto \text{prob}[D|S,E] \times \text{prob}[S|E] \, ,$$

where S is the scattering law, D are the data and E is the experimental setup. The term on the far right is called the *prior* probability distribution function (p.d.f.) and represents our state-of-knowledge, or the lack thereof, about the scattering law before we have any data. Its conditioning on the experimental setup can be dropped since the instrumental design is irrelevant if we do not have any data: prob[S|E] = prob[S]. Entropy, for example, enters the data analysis (or inference) through this term and incorporates the prior knowledge that the scattering law is a positive and additive distribution. Bayes' theorem tells us that our prior state-of-knowledge about the scattering law is modified by the data through the so-called *likelihood function*: prob[D|S,E]. It represents how likely it would be that we would measure the data that we did, given a (trial) scattering law and the experimental setup; it is often of the form $\exp(-\chi^2/2)$, where χ^2 is the usual sum-of-squared-residuals misfit statistic. The product of the prior p.d.f. and the likelihood function yields the *posterior* p.d.f. we require, representing our state-of-knowledge about the scattering after we have measured the data.

Since the instrumental design (only) enters the posterior p.d.f. through the likelihood function, we need to look at its sharpness or spread. We should optimise our instrumental design by making the likelihood function as sharply-peaked as possible. This prescription is quite general and applies to questions of experimental design in any context: (i) state the quantity of interest which you wish to infer; (ii) state the experimental design parameters at your disposal; (iii) see how the likelihood function, viewed in the space of the quantity of interest, changes with the experimental design parameters; (iv) optimise by making the likelihood function as sharply-peaked as possible. The sharper the likelihood function, the greater the sensitivity of the data to changes in the quantity of interest and, hence, the stronger the constraint imposed by the data on what the (value of) the quantity of interest could be.

Although this optimisation procedure is quite general, we now consider some specific examples relevant to optimising the instrumental resolution function for a neutron scattering instrument. Along the way, we will also make certain assumptions and approximations (usually valid) in order to obtain analytic solutions.

4. Examples

Before we can start any data analysis, and consider the likelihood function or the prior p.d.f., we must formulate the precise question we wish to answer. Formally, we must define the space of possible answers or choose the *hypothesis space*. In neutron scattering we may say that we wish to know the scattering law of our sample, but how is the scattering law to be described? If we know (or assume) that scattering law consists of a single Lorentzian, for example, then we have a 3-dimensional hypothesis space defined by the position, height and width of the Lorentzian. If, on the other hand, we have no functional form for the scattering law, then we might digitise it into large number of M pixels, where upon we have an M-dimensional hypothesis space defined by the flux in each pixel. The problem is that answer to the question "what is our best estimate of the scattering law" depends not only on the data but also on the precise formulation of the question, or our choice of hypothesis space. This is the source of our most basic difficulty in not being able to provide a universal "figure-of-merit" for instrument design. So, we will investigate three specific cases.

4.1 The simplest case

Let us begin by considering a very simple situation. Suppose we know that the scattering law consists of a single δ-function excitation of known magnitude but unknown position — i.e. we have a 1-dimensional hypothesis space defined by the position of the δ-function, x_0 say. Suppose also that the experimental data are the result of a convolution between this scattering law and a Gaussian resolution function $T.\exp(-x^2/2w^2)$. The height T of this Gaussian resolution function is determined by the length of time for which the data are collected, and its width w is some function of the instrumental parameters like flight-path length and collimation angle. The question now is: "what restriction do the data impose on the value of x_0?" The width of the likelihood bump, viewed in the 1-dimensional space of x_0, gives us the uncertainty in the position δx_0 allowed by the data. After some algebra, we find that δx_0 depends on the the the instrumental design, enshrined in the resolution parameters T & w, in the following manner: $\langle \delta x_0^2 \rangle \propto w/T$. The inverse of this quantity can be used as a figure-of-merit, and has been quoted in the neutron scattering literature (Michaudon 1963; Day & Sinclair 1969; Windsor 1981):

$$\text{Conventional figure-of-merit} = \frac{\text{Total number of neutrons}}{(\text{FWHM})^2} \propto \frac{T}{w} \, ,$$

where FWHM is the full-width-half-maximum of the resolution function, and the total number of neutrons detected is proportional to Tw.

We now give a couple of simulated examples to make the point that, although this figure-of-merit is the correct answer to the question posed above (i.e. what is the uncertainty in the value of x_0?), it is quite unsuitable for general use; for example, it would be a bad guide for judging our ability to resolve closely-spaced peaks or for inferring their widths & magnitudes. The test object ("typical" scattering law) of Fig. 3(a) was convolved, independently, with the two resolution functions shown in Fig. 3(b) to generate two noisy data-sets similar to those shown in Figs. 2(b) & 2(c); the MaxEnt reconstructions derived from these data-sets are given in Fig. 3(c). Even though the resolution functions had identical figures-of-merit according to the equation above, the reconstruction from the sharp-edged resolution function is clearly far superior to the other. We emphasise that the use of MaxEnt to carry out the deconvolution (strictly speaking, the use of an entropic prior to carry out the inference) is not central to the point we are illustrating — the difference in the quality of the reconstructions is a reflection on the nature of the data, and not on the properties of MaxEnt. But the figure-of-merit above was based on a Gaussian resolution function, you might complain, and so is not valid here. In Fig. 4 we show the result of applying the same test to data from two resolution functions which are both Gaussian but with a FWHM ratio of 10:1. Again, according to conventional thinking, the figures-of-merit can be equalised by increasing the total number of counts for the wide

Gaussian by a factor of 100. Fig. 4 illustrates the point that this is not the case in general — to recover the sharpest structure with the wide resolution function one would need to increase the number of neutrons counts by many orders of magnitude!

4.2   The second most simple case

Next, let us move on to consider a slightly more complicated case. Let the situation be exactly the same as before, except that now the scattering law is known to consist of a single δ-function of not only unknown position but also unknown magnitude — i.e. we have a 2-dimensional hypothesis space, defined by the magnitude A and position x_0 of the δ-function. Again, we need to ask the question: "what restrictions do the data impose on the value of A and x_0?" The likelihood function is now a bump in a 2-dimensional space, a general schematic illustration being given in Fig. 5. To describe the shape of this probability bubble, we need at least three numbers: two for the widths, in each of the 2 dimensions, and one for the orientation. One way of doing this is to give the so-called *covariance matrix*. The elements of this 2x2, symmetric, matrix tell us the expected uncertainty in the position $\langle\delta x_0^2\rangle$, the expected uncertainty in the magnitude $\langle\delta A^2\rangle$, and how uncertainties in one affect the other $\langle\delta x_0\delta A\rangle$. After some algebra, we find that the correlation term $\langle\delta x_0\delta A\rangle=0$ indicating that the reliability with which we can estimate the position of the δ-function has no bearing the reliability with which we can estimate its magnitude; in terms of the general schematic picture

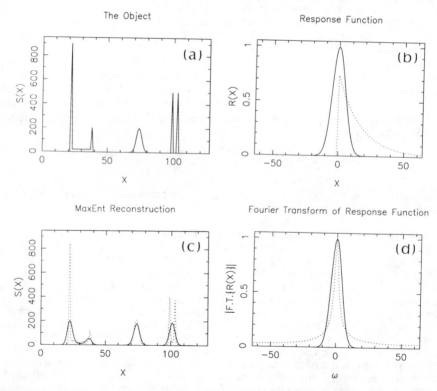

Figure 3: **(a)** The test object, or "typical" scattering law, used to generate simulated data. **(b)** Two resolution functions having identical conventional figure-of-merits. **(c)** The MaxEnt reconstructions obtained from noisy data generated by the convolution of (a) with the resolution functions in (b). **(d)** Fourier transforms of the resolution functions in (b).

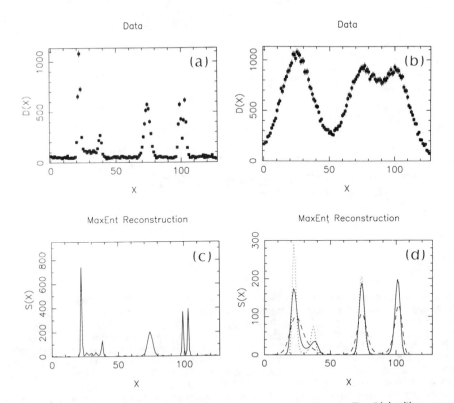

Figure 4: **(a)** Noisy data resulting from the convolution of the object in Fig. 3(a) with a narrow Gaussian. **(b)** Corresponding data for a wide Gaussian. **(c)** MaxEnt reconstruction from the data in (a). **(d)** The dashed line is the MaxEnt reconstruction from the data in (b); the solid line is from data with 100 times, and the dotted line 10000 times, the number of counts as in (b).

of Fig. 5, this means that the principal axes of the likelihood probability bubble should lie along the A and x_0 directions. We also find that instrumental design parameters, inherent in T and w, affect the reliability of the inferred magnitude and position of the δ-function as follows:

$$<\delta x_0^2> \propto w/T \quad \text{and} \quad <\delta A^2> \propto 1/Tw .$$

This illustrates another fundamental dilemma for instrument design by raising the question: "what do we mean by a figure-of-merit?" The formulae above say that to improve our estimate of the position of the δ-function, we should make the width of the Gaussian resolution function as narrow as possible; to improve our estimate of its magnitude, we should make the resolution function as wide as possible!

4.3 A generalised case

We can, of course, keep working through specific problems, but we will only come up with the conclusion that "different questions, or choice of hypothesis spaces, have different answers". So, let us try to ask a *generalised question*; we accept that it will not give the exact answer to every specific case, but hope that it will yield a sensible figure-of-merit for a wide range of situations.

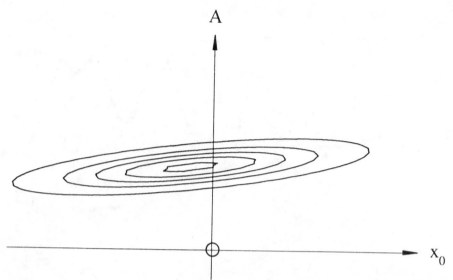

Figure 5: A schematic diagram of a 2-d likelihood function: pr({Data}|A,x_0). The horizontal and vertical axes represent the values of the two parameters, A & x_0, that we wish to estimate; the contours are lines of equal likelihood. The shape of the likelihood function can be described either by the *covariance matrix* or by the *eigenvectors* & *eigenvalues* of the (logarithm of the) likelihood function. The elements of the covariance matrix tells us the expected uncertainties allowed by the data in our estimates of A & x_0, $<\delta A^2>$ & $<\delta x_0^2>$, and how our estimate of one affects the other $<\delta A \delta x_0>$. The eignvectors are the directions of the principal axes of the likelihood "bubble", and the eigenvalues are related to the widths of the likelihood function in these directions. They tells us which properties, or linear combinations, of A & x_0 can be determined independently of each other, and how reliably they can be estimated.

Let us say that the instrumental parameters like moderator material, moderator temperature, flight-path length, collimation angle, and so on, all combine to give some resolution function R(x) (not necessarily Gaussian). The question we will ask is: "Given that the data are the result of a convolution between the sample scattering law S(x) and the instrumental resolution function R(x), how reliably can we estimate the scattering law assuming no particular functional form for S(x) ?"

Since we do not have a functional form for the scattering law, as we did before, an obvious hypothesis space to choose is the one defined by the values of S(x) specified on a finely digitised grid in x — i.e. we have a very large dimensional hypothesis space. The likelihood function will now be a bump in this large multi-dimensional space, and we can consider Fig. 5 as a schematic 2-dimensional slice through it except that the axes labels should now read $S(x_i)$ & $S(x_j)$ instead of A and x_0. The spread of this multi-dimensional probability bubble about its maximum will, of course, give us a measure of how well the data constrain the permissible scattering laws. Since this likelihood bubble will, in general, be skew with respect to our $\{S(x_j)\}$ coordinates, it is difficult to describe its width. It is convenient, therefore, to rotate our coordinate axes from the original $\{S(x_j)\}$ to another set which lie along the principal axes of the probability bump; the spread of the bubble is then just given by its width along these new coordinate axes. These principal axes are vectors in the coordinates $\{S(x_j)\}$ and hence represent relative pixel heights in our digitised x-coordinate — they are discretised functions of x. Formally, the principal axes are the *eigenvectors* , or *eigenfunctions* if we go to the continuum limit, of the logarithm of the likelihood function.

The eigenfunctions define the natural hypothesis space for our problem because they represent the properties of the scattering law which can be estimated independently of each other. If we write the scattering law as a linear combination of these eigenfunctions, $S(x) = \sum a_j.\eta_j(x)$, where $\eta_j(x)$ are the eigenfunctions, and a_j are coefficients (or parameters) which are now be determined from the data, then we will find that our reliability in the estimate of one parameter will not affect our estimate of another — i.e. the covariance matrix is diagonal ($<\delta a_i \delta a_j>=0$). The widths of the likelihood function along these principal directions tells us the reliability with which the eigenfunction properties of the scattering law can be estimated; the widths are related to the eigenvalues λ by: $<\delta a_j^2> = 2/\lambda_j$.

If we were to carry out the algebra for our problem (Sivia 1990), making suitable (usually reasonable) assumptions to obtain an analytic solution, we would find that the eigenfunctions and their corresponding eigenvalues, labeled by ω, are given by:

Eigenfunctions: $\quad \eta_\omega(x) = Cos(\omega x) \ \& \ Sin(\omega x)$, with

Eigenvalues: $\quad \lambda_\omega \quad = \dfrac{2}{\sigma^2}.|\tilde{R}(\omega)|^2$.

where $\tilde{R}(\omega)$ is the Fourier transform of the resolution function $R(x)$, and σ^2 is a measure of the average number of counts in the data. This tells us that if we do not have a functional form for the scattering law, then we should express it in terms of a Fourier series (the sum of sine and cosine functions). The advantage of doing this is that reliability with which we can for the scattering law, then we should express it in terms of a Fourier series (the sum of sine and cosine functions). The advantage of doing this is that reliability with which we can estimate one Fourier coefficient will not affect the accuracy with which we can determine another — it is an uncorrelated space. Since the reliability with which we can estimate any Fourier coefficient is related to the reciprocal of the corresponding eigenvalue, we can use λ_ω as a figure-of-merit for inferring structure in the scattering law with detail $\Delta x \approx 1/\omega$.

5. Conclusions

The implications of this analysis for a figure-of-merit for instrument design are as follows:

(a) A versatile figure-of-merit depends largely on the Fourier transform of the resolution function rather than on its full-width-half-maximum. We can see this illustrated in Fig. 3: the two resolution function in Fig. 3(b) have the same full-width-half-maximum, and integrated intensity, but the Fourier transform of the one with the sharp edge does not decay as rapidly with increasing frequency ω as the the one from the Gaussian, as seen in Fig. 3(d). Resolution functions which have sharp features, therefore, allow high-frequency information to be recovered reliably from the data. In electrical engineering we would say that the figure-of-merit is governed by the *bandwidth* of the resolution function.

(b) The figure-of-merit is not constant for a given resolution function, but depends on the amount of detail required in the inferred scattering law.

(c) The background signal has not been forgotten, and enters the figure-of-merit through the average number of counts, or σ^2, dependence. Any long decaying tail of the resolution function reduces the figure-of-merit in the same way as the background since it adds to the average number of counts but does not contribute to the Fourier term $\tilde{R}(\omega)$ at high frequency.

Since the resolution function in neutron scattering depends on the details of the spectrometer & moderator, there is potential for a revision of ideas on the design of neutron scattering facilities. Take, for example, the *matching* of resolution elements on a neutron spectrometer at an acclerator-based source, which is illustrated in Fig. 6. The resolution function for an

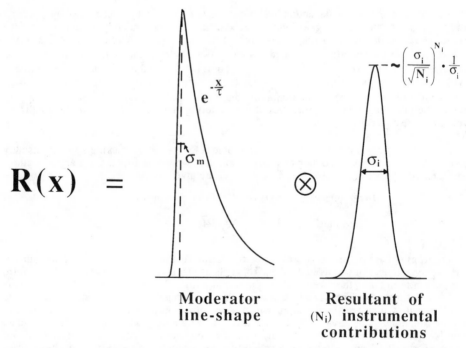

Figure 6: A schematic picture of resolution *matching* at a pulsed neutron source. instrument design questions include: what is the optimal value of σ_i (i.e. flight-path, collimation, etc.) given σ_m & τ (the moderator)? What is the best choice of σ_m (e.g. moderator material) and τ (poison)?

experiment is the resultant of a convolution between the moderator pulse-shape, which has a sharp rising edge and a long decaying tail, and a roughly a Gaussian component from the instrumental contributions (e.g. flight-path, collimation, etc.). The question is how to choose the width of the instrumental component so as to get the "best" resultant resolution function. Conventional wisdom recommends that we should make the width of the instrumental, Gaussian-like, contribution comparable to the width of the moderator pulse-shape. The analysis above, however, suggests that following this advice could seriously impair our ability to infer (reliably) the scattering law at high resolution and that what we should probably do is to match to the narrow width of the sharp leading edge. How this translates into the optimal choice for collimation and flight-path length, or the moderator material chosen to control the sharpness of the leading edge, or whether we should "poison" the moderator to reduce the decaying tail, are the subject of on-going research.

Finally, let us return to the example of Fig. 2. We said in Section 2 that a conventional optimisation, aided by visual intuition, would lead us to believe that the data of Fig. 2(a) were far superior to those in Figs. 2(b) & 2(c) and that the data of Fig. 2(b) were slightly more preferable to those of Fig. 2(c). An analysis based on the Bayesian approach presented here, however, would suggest that the quality of the data in Figs. 2(a) & (c) was comparable and far superior to that of Fig. 2(b). The MaxEnt reconstructions from these three data-sets, shown in Fig. 7, illustrate graphically that the latter statistical inference approach to optimisation is much to be preferred over traditionally used procedures.

Acknowledgements

This work was supported by the Office of Basic Sciences of the U.S. Department of Energy.

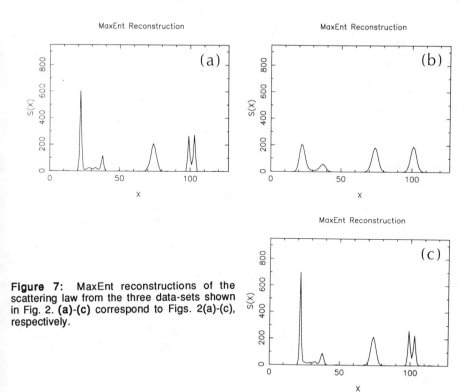

Figure 7: MaxEnt reconstructions of the scattering law from the three data-sets shown in Fig. 2. **(a)-(c)** correspond to Figs. 2(a)-(c), respectively.

References

Day, D.H. & Sinclair, R.N. (1969). *Neutron Moderator Assemblies for Pulsed Thermal Neutron Time-of-Flight Experiments*, Nucl. Instr. Meth., 72, 237-253.

Jeffreys, H. (1939). *Theory of Probability*, Oxford University Press. Fourth edition: 1983.

Jaynes, E.T. (1986). *Bayesian Methods - an Introductory Tutorial*. In Maximum Entropy and Bayesian Methods in Applied Statistics, J.H. Justice ed., Cambridge University Press.

Michaudon, A (1963). *Reactor Science and Technology*, Journal of Nuclear Energy A/B, 17, 165-186.

Silver, R.N., Sivia, D.S., Pynn, R. (1989). *Information Content of Lineshapes*, Advanced Neutron Sources 1988, D.K. Hyer ed., Institute of Physics Conf. Series 97, 673-683.

Sivia, D.S., Silver, R.N., Pynn, R. (1990). *Optimization of Resolution Functions for Neutron Scattering*, Nucl. Instr. Meth., A287, 538-550.

Sivia, D.S. (1990). *Applications of Maximum Entropy and Bayesian Methods in Neutron Scattering*, in Maximum Entropy and Bayesian Methods: Dartmouth 1989, P. Fougere ed., Kluwer Academic Publishers.

Windsor, C. (1981). *Pulsed Neutron Scattering*, Taylor & Francis Ltd., London.

An amateur's guide to the pitfalls of maximum entropy

A.K.Soper

Neutron Science Division,Rutherford Appleton Laboratory, Chilton, Didcot,Oxon,UK

Abstract. Maximum entropy (ME) techniques are being increasingly used to perform linear inversions of experimental data. The method allows the noise and resolution broadening to be taken account of in a quantitative manner. Questions must be raised however about what is the true information content of a probability distribution. For Fourier transforms and deconvolution it is well known that ME reconstructions often have the same "ringing" seen in direct methods. It is found here that a restraining function based on the first and second derivatives of the distribution gives excellent results in both cases.

1. INTRODUCTION

The interpretaton of most experiments is affected by the finite accuracy of measurements. Even if there were no other limitations measured quantities can rarely be compared directly with theory, but instead via some transformation of the theoretical function. There are additional complications such as the finite resolution of the measuring apparatus, statistical noise and other systematic errors. The data themselves are frequently incomplete on account of the type of apparatus being used or the nature of the experiment.

In this situation there are two ways to proceed. On the one hand the data can be compared with theory by convoluting the theoretical function with all the known experimental artifacts and so comparing in data space. Frequently the procedure is executed iteratively with a gradual refinement of the model until a satisfactory fit of model to data is achieved.

Such a procedure however may not work if the initial guess for the model is too far away from the real data, and moreover there is little indication of the range of model predictions which might equally be regarded as compatible with the data. The alternative to this indirect approach therefore is to attempt to generate the theoretical function by a direct inversion of the data. Because of the data's limitations this is usually impossible without the application of a constraint additional to the simple one of requiring a fit to the data.

It is precisely the lack of uniqueness in the inversion of experimental data which has prompted the search for an underlying "quality factor" to select, out of the vast array of compatible results, those which are most likely to be correct. Shannon and Weaver (1949), Jaynes (1982), and others have argued that the natural choice for such a quality factor is the entropy, H, which measures the extent of the disorder in any particular distribution: the most likely distribution is the one that makes the least commitment to the missing data. In other words of all possible distributions the most likely one is also the most disordered one. Clearly in the absence of any data all distributions are equally likely so there is complete disorder. Formally entropy is defined as

$$H = p_k ln(p_k/b_k) + p_k - b_k \tag{1}$$

where p_k is a distribution function, and b_k represents prior knowledge about the distribution, i.e. those features which it is known p_k must satisfy. This definition is derived from quite fundamental arguments such as occur for example in statistical thermodynamics. Where the controversy arises apparently is in the application of this formula: exactly what is the distribution function whose entropy we are to maximize and what is the prior information that is required? In practice the question of what prior information to introduce is subjective and different choices can lead to quite different conclusions.

2. THE MONTE CARLO APPROACH

In the Monte Carlo (MC) approach it is assumed that there exists an ensemble of distributions, $N_{j,k}$, and the k'th distribution occurs with probability p_k, with prior probabilities in b_k. The entropy of the ensemble (1) has to be maximised. If there are no data and no prior information then $p_1 = p_2 = \cdots = p_k = \cdots$, i.e. any distribution can occur with equal probability. If data are present the k'th distribution produces a "model" of the data, $M_{i,k}$, which is compared to the data via χ^2:

$$\chi_k^2 = \sum_i \frac{(D_i - M_{i,k})^2}{\sigma_i^2} \tag{2}$$

which strictly has to be constrained to the number of data points. The solution to this problem is simply

$$p_k = b_k exp(-\lambda\chi^2) \tag{3}$$

where λ is an undetermined multiplier which controls how well the average distribution approaches the data. This therefore is the most reasonable way of assigning probabilities to a set of distribution functions, and forms the basis for the Monte Carlo algorithm described in this paper. The technique is also sometimes called *simulated annealling* (Pannetier (1990)) or *Reverse Monte Carlo* (McGreevy, 1990) because of the similarity of the approach to the classical Monte Carlo method used in condensed matter studies. It is however not a widely used method in the context of data analysis since it still requires a search of all possible distributions, a process which can be time consuming if there really is no prior information on the form the distribution should take. One of the frequently quoted advantages of the Monte Carlo approach is that moves in which χ^2 gets worse are often accepted so the simulation is less likely to get stuck in local "fals" minima than other direct minimization techniques.

3. THE MAXIMUM ENTROPY APPROACH

In the Maximum Entropy (ME) approach instead of assigning some overall probability to the entire distribution N_j as in the Monte Carlo approach, a separate probability p_j is associated with each pixel, N_j. In fact it is usual to write

$$p_j = aN_j, \quad b_j = aB_j \tag{4}$$

where B_j is the prior knowledge of N_j, i.e. there is an assumed linear relationship between p_j and N_j. There is no particular justification for this step, although it is not unreasonable - if there were no data then the p_j's would be uniform and so therefore would the N_j's. However it does place a restriction on the N_j in that it implies they are normalizeable, and also that they do not become negative. Neither restriction is necessarily true in general.

The ME solution for p_j and therefore N_j in the case of a linear transform is

$$N_j = B_j exp\left[-\lambda \sum_i \frac{2(D_i - M_{i,k})}{\sigma_i^2} T_{i,j}\right] \quad (5)$$

where $T_{i,j}$ is the transform matrix from N_j to M_i:

$$M_i = \sum_j T_{i,j} N_j \quad (6)$$

Although equation (5) is highly non-linear a solution can be found by standard search procedures, Bryan and Skilling (1984). It can also be solved by the Monte Carlo technique as shown below. Again λ is the undetermined multiplier which controls how well the χ^2 constraint is satisfied.

4. COMPARISON OF MONTE CARLO AND MAXIMUM ENTROPY

In the Monte Carlo approach, in order to proceed on a random walk through configuration space it is necessary to calculate the transitional probability

$$\Delta p_k = p_k(N_j + \Delta_j) - p_k(N_j) = p_k(N_j) \, exp(-\Delta\chi^2) \quad (7)$$

where $\Delta\chi^2$ is the change in χ^2 with the change $N_j \to N_j + \Delta_j$. It is straightforward to show that

$$\Delta\chi^2 = \left[\sum_i \frac{2(D_i - M_{i,k})}{\sigma_i^2} T_{i,j}\right]*\Delta_j + \left[\sum_i T_{i,j}^2\right]*\Delta_j^2 \quad (8)$$

Comparison with (5) shows that the transitional probability in the MC approach looks to first order very similar to the solution for N_j in the second. The differences are that there are now no restrictions on the N_j being always positive (although positivity can be enforced if needed), and that there is a quadratic term in the exponent, which serves to ensure that large fluctuations in N_j are rejected.

The Monte Carlo approach yields an ensemble of solutions. The final answer is then the ensemble average of all the particular solutions:

$$\overline{N}_j = \sum_k p_k N_{j,k} \quad (9)$$

with variance

$$\epsilon_j^2 = \sum_k p_k(N_{j,k} - \overline{N}_j)^2 \quad (10)$$

Hence a realistic uncertainty in the simulated result is obtained assuming of course the simulation has proceeded on a true random walk through configuration space. Recent versions of ME are apparently doing an equivalent calculation (Gull and Skilling, 1989) although traditionally ME has yielded only a single solution.

The standard ME approach can also be solved by the MC method by noting that equation (5) is a result of minimizing the function:

$$F = \lambda\chi^2 - H \tag{11}$$

Therefore this problem can be solved by replacing $\Delta\chi^2$ by ΔF in (7), but otherwise proceeding as before, i.e. performing a Monte Carlo simulation with transition probabilities based on the value of F rather than χ^2 alone.

When written like this however it can be seen that that the real rôle of entropy in ME is simply that of a restraining function on the N_j's. Since H is maximized when $N_j = B_j$ for all j, H has the effect of driving the distribution towards the prior in preference to other directions that it might wish to go, e.g. as directed by the χ^2 constraint. The final solution is then a balance between these two often opposing constraints. What is not established however is that H is the correct restraining function to use in every case: in the MC method use of equation (3) with a uniform prior is very reasonable since there is no "a priori" reason to expect one ensemble to be connected in any way with the next. But in ME using equation (4) it is UNLIKELY that the N_j would be uncorrelated, so additional information is needed in the form of a prior. Such information is not necessarily well defined.

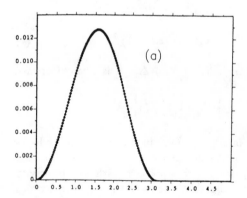 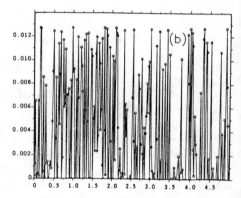

Figure 1. Comparison of two distributions with the same entropy according to equation (1). Distribution (b) on the right is a random re-ordering of the points in (a).

Consider for example the case of the transform of liquid structure factor data, S(Q), to pair correlation function, g(r). In this case the normal choices for N_j and B_j are

$$N_j = 4\pi\rho r_j^2(g(r_j) - 1)\Delta r_j, \quad B_j = 4\pi\rho r_j^2\Delta r_j, \tag{12}$$

where ρ is the density of the fluid and r_j is the distance from an atom at the origin. Note that with this choice the N_j are normalizeable, but they also take on both positive and negative values. The corresponding transform matrix is

$$T_{i,j} = \frac{\sin Q_i r_j}{Q_i, r_j} \tag{13}$$

where Q_i is the wavevector transfer in the diffraction experiment.

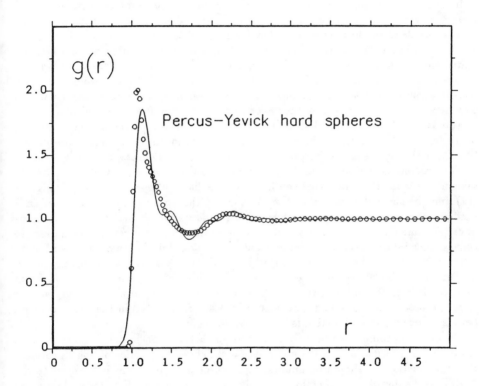

Figure 2. ME reconstruction of the Percus-Yevick hard sphere pair correlation function derived from a truncated structure factor (—) compared to that obtained by minimizing the noise function (14) (o o o).

When substituted into (5) or (8) it can be seen that the exponent is in fact a (modified) Fourier transform of the difference between the data and the model and over a finite range of Q_i values, so that the exponent will be biassed by both the truncation and the noise in the data. If the chosen prior does not specifically reflect that bias there is nothing to guarantee that those effects will not be reproduced in the final solution. Using the standard prescription of using a uniform prior it is not clear that the ME solution has in fact gained us very much over a direct Fourier inversion of the data. What we have done however is to transfer the problem of obtaining a "reasonable" set of N_j's to one of defining "reasonable" prior information, B_j. If the choice of prior is done by guesswork or by trial and error, then it is likely that the ME reconstruction will be as arbitrary as the solution obtained by other means. Clearly therefore the key to a successful reconstruction lies in developing a prior distribution which accurately combats those artifacts which we KNOW to be present in the data, but which imposes no other constraints on the solution.

5. THE MINIMUM NOISE APPROACH

In order to develop a practical approach to the problems of deconvolution and Fourier inversion an heuristic argument is used here. There is an implicit assumption in most problems of physical interest that any given distribution function is continuous and has continuous derivatives. For example in quantum mechanics since physical potential energies are likely to be continuous, however sharply varying, the wave function is expected to be continuous. Without this assumption no solutions would be possible. In experimental science although data are measured at discrete points and can only be stored in discrete pixels, the underlying distribution functions which generated those data is always assumed to be everywhere continuous. Even Bragg diffraction peaks from real crystals, which would be δ-functions in an ideal crystal, are in fact broadened, however slightly, by mosaic spread, particle size effects, defects, and so on.

The problem is that as it stands ME with a uniform prior does NOT automatically take into account the underlying continuity of the distributions being sought. Figures 1(a) and (b) show the same set of N_j but in (b) they have been redistributed in a random manner. With no prior information (B_j is uniform) these two distributions have the same entropy, if entropy is defined according to (1), yet curve (a) appears to convey some information, (e.g. the centroid and standard deviation of the distribution relate definitely to what can be seen), while for (b) there is little apparent significant information. It is claimed that the data will supply the necessary correlations which will serve to distinguish between (a) and (b). However this still is not a satisfactory answer since the data may well contain artifacts which are reproduced in the reconstruction because the restraining function is insensitive to them.

Of course continuity can be introduced into the ME approach, but it has to be done via the prior information. In the absence of well defined prior information I postulate that the function we should be minimizing in all of these problems is sensitive to the NOISE in the distribution function: the solution to a fitting problem most likely to be correct is also expected to be the LEAST NOISY. This is distinct from the standard ME approach which attempts to make the final reconstruction most UNIFORM. Since the data exist in the first place it is UNLIKELY that the correct solution is uniform!

To quantify noise the function used here relies on the fact that the derivatives of a function exaggerate the noise in that function. A practical function which gives excellent results is generated from the first and second derivatives of the distribution function:

$$I_k = \sum_j \frac{(N_{j,k} - B_{j,k})^2}{|N_{j,k-1} - N_{j-1,k-1}| + |N_{j+1,k-1} - N_{j,k-1}|} \tag{14}$$

where the prior, $B_{j,k}$, is defined by

$$B_{j,k} = \frac{1}{2}(N_{j-1,k-1} + N_{j+1,k-1}) \tag{15}$$

The top line of the function I_k, which is essentially the square of local second derivative of the N_j's, is sensitive to the noise in the distribution, while the denominator is

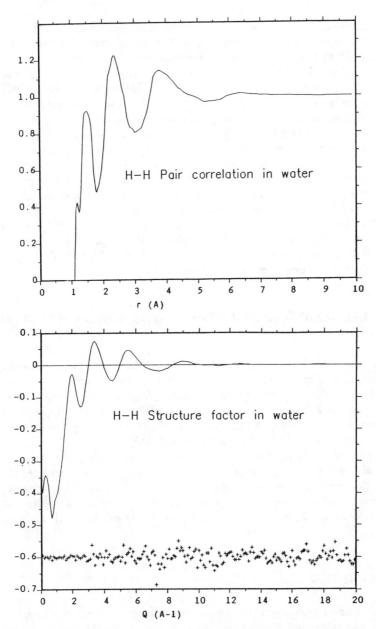

Figure 3. Minimum noise reconstruction (top) of the H-H correlation function in liquid water from noisy structure factor data. Below is shown the reconstructed structure factor and the residue between fit and data.

a weighting function which serves to de-emphasize the importance of noise in those regions where N_j is changing rapidly. Hence this function has the important attribute that it allows for rapid changes in N_j with j only if they can be justified by the data, otherwise it tries to make the N_j as smooth as possible. The noise function (14) probably could be cast in the form of an entropy function like (1) but there is little reason to do so: it should only be regarded as a practical restraining function on the N_j, which is fulfilling the same rôle that the entropy function does in the standard ME method. The difference is that (14) assumes the distribution to be highly correlated unless the data are strong enough to reduce that correlation, while standard ME with uniform prior assumes the distribution is uncorrelated unless data require otherwise.

There is no claim that (14) is the only function of its kind. However what is clear is that minimizing the NOISE is the the key to producing a reliable reconstruction. If the driving function in (8) tries to put in a peak into the reconstruction, then the noise function will tend to oppose that peak. On the other hand if several neighbouring N_j are correlated around a peak then the noise function will be less resistive. This function combined with the Monte Carlo sampling technique has provided consistently good Fourier transforms, including the case of a truncated hard sphere structure factor (see figure 2), a problem which was addressed at some length at the previous WONSDA meeting (Root, Egelstaff and Nickel 1986), or when the original data are very noisy (figure 3). It has also been successfully applied to the problem of deconvoluting a complex resolution function from experimental data.

6. AN EXAMPLE: DECONVOLUTING THE PHONON WINGS FROM TFXA DATA

The Time-Focussed Crystal Anlayser spectrometer (TFXA) at ISIS produces a spectrum of inter- and intra-molecular vibrational and rotational excitations for a given material. Because the wavevector transfer on this instrument is significant and dependent on the energy transfer the intrinsic vibrational spectrum of the material is broadened by phonon excitations and the shape of the broadening function is dependent on the energy transfer (Tomkinson and Kearly 1989).

The method has clearly established all the principal modes, plus indicated two overtone modes which are not shown in the calculation. It has also revealed the time structure in the neutron pulse, as seen from the assymmetry in the deconvoluted peaks, which was not included in the peak broadening function of figure 4(a). Note also there is apparently little of the ringing which is frequently associated with deconvolution methods.

7. CONCLUSION

The Monte Carlo simulation method with a restraining function based on minimizing the noise in the calculated distribution function, equation (14), has now been used to analyse a large number of Fourier inversion problems for data taken on liquids, amorphous solids and crystalline powders. In every case the results are consistently superior to the method of direct Fourier inversion. As seen above the method can also be used for reliably deconvoluting resolution functions without the need for specifying prior information. Because there is no reliance on establishing a prior distribution, the results are not dependent on the choice of starting configuration. Earlier accounts of the present approach have already appeared (Soper 1989) but the latest version has been found to be superior. The principle enhancement over the previous versions is

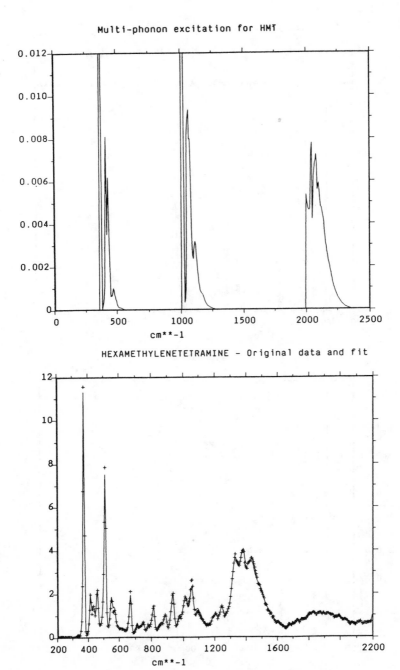

Figure 4. (top) The estimated broadening function for a molecular solid, hexamethylenetetramine iodide (HMT) at three energy transfers. The data to be analysed and the minimum noise fit are shown below.

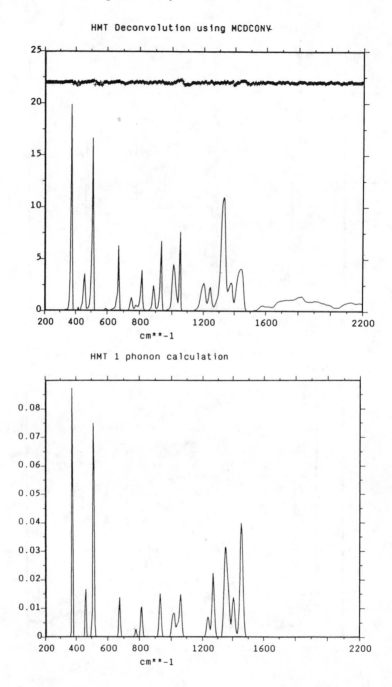

Figure 5. Result of the minimum noise deconvolution (top), compared with a normal mode calculation (Jobic and Lauter 1988) (bottom).

that there is only one parameter to specify, i.e. the level of fit to the data. If the statistical error bars were the only source of error then the fit would be defined by χ^2, but in practice the process of recording the data and subsequent correction procedures can introduce significant systematic effects. For the g(r) problem these can be partly accounted for by requiring g(r) to be zero over a region of r.

It is unlikely that the story is complete yet. The noise function (14) is sensitive to the obvious high frequency artifacts in inversion problems, such as noise and truncation errors. However very often in Fourier transforms there is missing information at low frequencies as well, and the present noise function is rather insensitive to low frequency artifacts: they invariably involve correlations between a large number of N_j and so appear as valid data. Further work on this problem is needed.

Acknowledgments

I would like to thank Bill David and Mike Johnson for several illuminating discussions on the material for this paper. Also thanks to John Tomkinson for drawing my attention to the TFXA problem.

References

Bryan R K and Skilling J, 1984, Monthly Not. Astr. Soc.,**211**, 111

Gull S F and Skilling J, 1989, *Quantified Maximum Entropy "MEMSYS 3" Users' Manual*, Version 2.00, (Maximum Entropy Data Consultants, Royston, England).

Jaynes E T, 1982, Proc. of the IEEE., bf 70, 939

Jobic H and Lauter H J, 1988, J. Chem. Phys., bf 88, 5450

McGreevy R L, 1990, paper in this proceedings

Pannetier J, 1990, paper in this proceedings

Root J H, Egelstaff P A, and Nickel B G, 1986, *Neutron Scattering Data Analysis*, ed. M W Johnson, (IOP Conference Series no. 81, IOP Publishing , Bristol)

Shannon C E and Weaver W, 1949, *The Mathematical Theory of Communication*, (University of Illinois Press)

Soper A K, 1989, *Advanced Neutron Sources 1988*, ed. D K Hyer, (IOP Conference Series no. 97, IOP Publishing, Bristol) p711

Tomkinson J and G J Kearley, J. Chem. Phys., **91**, 5164

Inst. Phys. Conf. Ser. No 107: Chapter 1
Paper presented at Neutron Scatt. Data Anal. Conference, Rutherford Appleton, 1990

69

GENIE Version 3: A tool for neutron scattering data analysis and visualization

C M Moreton-Smith

Rutherford Appleton Laboratory, Chilton, Didcot, England OX11 OQX

Abstract: GENIE is a program written specifically for analysing and visualising the results of neutron scattering experiments. This paper describes briefly the limitations of the current version of the program (V2) in its intended role. It goes on to describe some current trends in scientific visualisation and analysis software in relation to the next version of the program GENIE Version 3. Finally, the operation of GENIE-V3 is discussed in more detail.

1. INTRODUCTION

Data analysis at the ISIS neutron source relies heavily on a program written for DEC VAX computers called GENIE. The GENIE program is used extensively in two ways; firstly as a spectrum visualization tool and secondly as a tool for data manipulation. The current version of the program, GENIE-V2, is limited in language and functionality and hence is restricted to preliminary data analysis and the manipulation of one-dimensional spectra.

The widespread use of the GENIE program in its current form is evidence that the original design (David et al 1986) targeted a major need in neutron Time Of Flight (TOF) data analysis. Since the original design in 1984 it has become clear that there is a wider need amongst the scientific community for powerful tools for data manipulation and visualization. These tools remove the need for scientists to waste time writing FORTRAN code which becomes duplicated in every establishment and often many times within the same establishment.

In the light of this need, GENIE-V3 is being written to encapsulate the ideas of GENIE-V2 while extending the functionality, firstly to rectify obvious deficiencies in GENIE-V2 and secondly to bring the program closer to the concept of a general "Scientist's work-bench." Extending this analogy, the "tool kit" will contain tools useful to neutron scatterers.

2. GENIE-V2 OVERVIEW

The first version of GENIE in general use on the ISIS instruments was substantially the same program as is in use today, the only changes since 1984 have been minor or peripheral such as the addition of extra graphics device drivers. Without doubt, GENIE-V2 has been a highly successful piece of software. The strengths of the program lie in several directions. A broad outline of the strengths and weaknesses is given below.

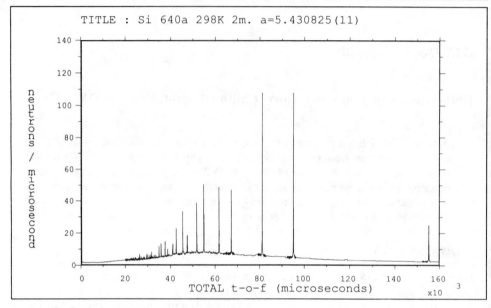

INSTRUMENT: HRPD   USER: DAVID/JOHNSON
RUN NUMBER: 93   RUN START TIME: 22-JUL-1985 17:02
SPECTRUM :   15   PLOT DATE: Tue 3-APR-1990 17:05
                                    BINNING IN GROUPS OF 10
LOCATION:

TITLE : Si 640a 298K 2m. a=5.430825(11)

Figure 1.

2.1 As a General Tool at ISIS

GENIE is used as a front end tool on all the ISIS instruments. This gives a uniformity to the preliminary data analysis which is much liked by the users. This has played a large part in the success of the program.

2.2 Access to data sources

A very strong feature of GENIE is its orthogonal access to data. The data may reside in files, the Front End Minicomputer (FEM) or the Data Acquisition electronics (DAE), in all cases the method of access is the same. Substituting the appropriate source in the "Assign" command selects the data for the subsequent "Display." For example,

            "ASSIGN DAE"
            "DISPLAY S1"

Selects the DAE as the source of the data and displays the first spectrum as a histogram. An identical sequence specifying a run number displays data from a raw data file instead (see Figure 1).

2.3 High Level Graphics

The single "Display" command shown in the above example is all that is necessary to draw a full histogram plot of the spectrum, again a very powerful feature, if necessary, other commands allow the basic display to be modified

2.4 Operations for TOF Neutron Scattering

Once the data source is selected, GENIE provides commands for preliminary analysis functions which can immediately be applied to this data, quite likely while an experiment is still in progress. For example, normalization and background subtraction can be carried out to ensure that the experiment is going correctly. The graphics system allows peak positions to be checked and plots to be annotated or over-plotted with control data for comparison.

2.5 Procedural Control

GENIE-V2 allows GENIE commands to be grouped as simple command procedures to automate analysis. Failings in the language at this level form one of the major deficiencies of GENIE-V2. For anything more than the simplest procedures the language is almost impossible to use.

2.6 Extensibility

"Hooks" are provided in GENIE-V2 to allow scientists to write FORTRAN programs to process data from GENIE and to return the results to GENIE. To some extent this alleviates problems with the paucity of the command language in the program itself.

3. Motivation for GENIE-V3

Although the problems with the command language are a serious deficiency in GENIE-V2, the reason for re-writing the program as GENIE-V3 is much more deeply seated than this. It is due to problems with the structure of the original FORTRAN code in which GENIE was written.

3.1 Alternatives to GENIE

At this point it is worth examining possible alternatives to GENIE altogether. Since GENIE-V2 was written, several broadly similar packages have become available which manipulate and display spectra; PV-Wave, a commercial package, has capabilities greatly exceeding those of the current version of GENIE, PAW (Physics Analysis Workstation) from CERN and the UNIMAP package from UNIRAS graphics are also contenders. It might seem pragmatic to dispense with GENIE at this point, scientists are quick learners and would rapidly learn to use a commercial package to advantage instead. The reason for keeping with GENIE is that unlike the alternatives, GENIE also embodies an intrinsic model of neutron scattering data. Looking more closely at some of the commercial packages. UNIMAP still does not provide a means of drawing error bars on its plots, let alone handling propagation of data errors. Even PV-wave, although providing a mechanism for drawing error bars provides no means of propagating errors through calculations. The deficiencies in both the commercial packages suggest that although these were written for scientists by technical people, practising scientists were not sufficiently consulted. The CERN package PAW was written for high energy physics data analysis, although a stronger contender from the scientific point of view, PAW still lacks the model of neutron scattering data.

3.2 Models

In the previous paragraph it was stated that GENIE contains a basic model of neutron scattering data. All computer programs model something. For the sake of terminology we shall name the thing being modelled an *entity,* usually something in the "real" world. Sometimes the entity being modelled is physical, at other times it is abstract . A lift control system models a physical entity - the lift system it controls. At other times the entity is abstract, many programs model an abstract entity such as an algorithm or process.

The basic entity which GENIE-V2 models is neutron scattering data represented as a TOF spectrum. The model is expressed both graphically (Figure 1) and textually (the commands used to produce Figure 1). By its success, it is reasonable to assume that GENIE-V2 has been a good model.

3.3 Representing the model in code

At the beginning of the section it was mentioned that there was a much deeper reason for re-writing the program than just to correct a few deficiencies. One of the essential requirements of a future GENIE is that it can be extended — or to be more precise — the set of models it supports can be extended. For example, GENIE-V2 has no concept of two-dimensional data and therefore cannot model the entity (Q,ω) space. To directly add this model to the GENIE-V2 code would be very awkward and might still fail to provide an adequate model.

The problem is that the monolithic code of GENIE-V2 *only* provides a model of TOF data as a one-dimensional histogram. It is quite easy to see how the entity being modelled "a one-dimensional histogram" can be generalized to "a two-dimensional histogram" (see Figure 2). Unfortunately, because the structure of the code in GENIE-V2 does not model the structure of the histogram entity, the generalization of the code that would be required is too major to be accommodated. All code wherever a histogram is used must be re-written.

There is a key assertion to be made here: *If the structure of the code follows the structure of the underlying scientific model, any logical refinement of the model will correspond to a logical refinement of the code.*

3.4 Structuring the code

Although the above assertion seems reasonable, it is not obvious how to make the structure of the code follow that of the entity being modelled. There are two ideas which help with this

Firstly, it follows inductively that if a complex physical entity can be simplified by breaking it down into components and the basic assertion above is true, then the code structure can also be simplified by this approach. In other words making the code model the structure of a complex entity is simply a matter of making the code model the structure of the simpler entities which comprise it.

Secondly, the model of an entity in the code must consist both of data structures and the functions which operate on them. Both should be localized in one area of the code and the data structures must not be accessed except through the functions provided.

Generalization of a 1-D Histogram to a 2-D Histogram

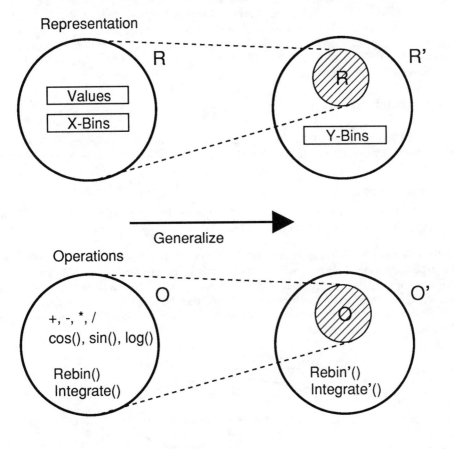

GENIE-V3   Cost of generalization = (R'-R)c + (O'-O)c

where c is a cost factor

GENIE-V2   cost as above + (O' x R)c

ie   All operations are likely to need modification not just those
      being implemented anew !

Figure 2.

In Figure 2 it can be seen that the cost of extending the code as the model is generalized depends mainly on the number of operations which will need to be changed if a particular data structure is changed. Where structures are accessed through functions as is proposed for GENIE-V3, the cost is the cost of implementing any additional functions (O'-O) and the cost of adding to the representation (R'-R). For GENIE-V2, also added to the cost is the need to add in a factor (O x R') where all *existing* operations must be changed to cope with the new representation as well.

4. GENIE-V3 Structure

Initially, GENIE-V3 does not aim to be too general in the models it represents. Here we look at how the structure of the code is built up to roughly the same level of model as embodied in GENIE-V2.

4.1 Entities modelled

Applying the ideas of the previous section, if we break down the entity TOF spectrum, one of the most basic entities is a number.

GENIE-V3 models real numbers to a high precision (usually double). As the numbers from real experiments are likely to be subject to uncertainty in measurement, GENIE-V3 can also model this uncertainty or error and propagate it through all calculations. Another feature of real-world experiments is numbers which are not there! These GENIE-V3 will treat as un-defined and propagate them correctly through calculations. Diagramatically the entity "Number" can be denoted thus.

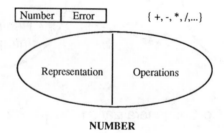

NUMBER

Number consists of a representation in the code and operations that act on that representation.

Now that the entity number has been defined, this can be used to build a vector. Time channels or data values may be modelled as vectors. Vectors in general may have any length, GENIE-V3 imposes no pre-defined length on vectors. Some operations though may require two vectors to be of identical length. As vectors consist of numbers, elements of vectors have all the features of numbers. Vectors now inherit features such as error propagation and undefined elements where required. This is clearly useful when taking the logarithm of a spectrum. Zero elements do not cause an error and are not erroneously set to zero or to some arbitrary default. The representation of vectors includes the representation of numbers and any operations on them as shown below.

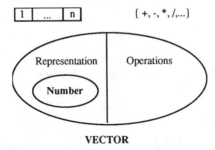

VECTOR

The concept can be further extended upwards by modelling the entity "Histogram" which makes use of two vectors in its representation.

Finally the entity TOF spectrum can be modelled by making use of the model of a histogram and including information such as units for the axes (which give the histogram meaning), time offsets and any other features specific to the TOF model.

Figure 3 shows in a simplified manner how the entities in GENIE-V3 are built up and how the complex models rely on those of the more basic entities.

5. Implementation of GENIE-V3

The implementation of GENIE-V3 presents several views of its internal models to the scientist using it (see Figure 4). Two of these are examined here. Firstly the textual or language view embodied in GENIE Command Language (GCL), and secondly the graphical view.

5.1 GCL implementation of the model

The mapping of the models described in the previous section onto a Command Language with a fixed syntax has to be done with great care. The danger is to naively identify the different models and then implement them as fixed types in the language with fixed functions which operate on them. The language would then support variables of many types directly derived from the abstract entities being modelled. This might seem ideal but the problem is that the models are rarely well enough defined unless they are very basic. GENIE-V3 makes a compromise, The types defined for GENIE-V3 are:

> **Number**
> Integer
> String
> **Numeric vector**
> Integer vector
> String vector
> **Workspace**
> **Workspace vector**

Entities modelled in GENIE-V3

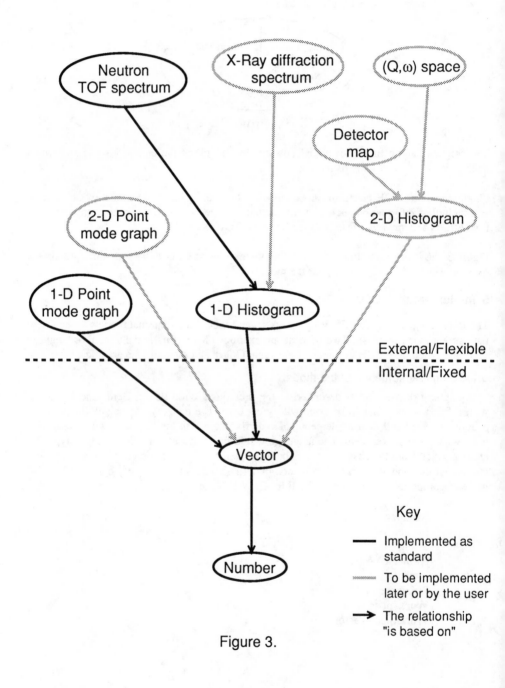

Figure 3.

External views of GENIE-V3 internal models

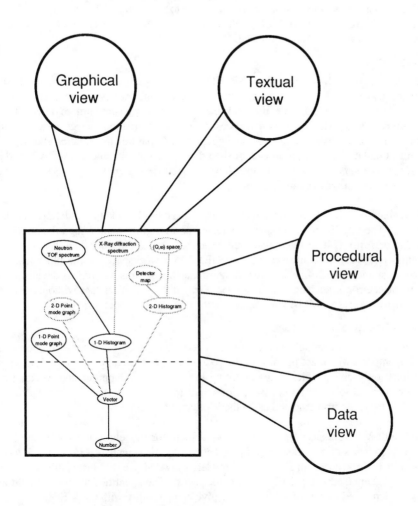

Figure 4.

Those types in bold are there primarily to support the scientific models. The other types are subsidiary and are only needed for operation of the language as a programming language where they provide facilities like loop counters (Integers) and axis labels (Strings). The obvious omissions from this list are types like "Histogram" and "TOF spectrum." The type "Workspace" also implies that GENIE-V3 is modelling some physical entity called a workspace which has not yet been mentioned.

5.2 Workspaces

In the early stages of the design of the GENIE-V3 language, there was much discussion about the concept of a "Workspace" arising out of ambiguities in the definition of the GENIE-V2 workspace. To some, the workspace was simply a spectrum, to others the essential feature of a GENIE-V2 workspace was the extra information carried around with the spectrum. In the discussion there was no agreement on what this information should be, each instrument had different requirements. It should be clear now that the GENIE-V2 workspace was acting as a general vehicle for modelling several different entities and as such was being viewed in different ways by different scientists.

In designing GENIE-V3, the concept of the workspace has been kept but defined much more carefully. The GENIE-V2 workspace contained no more information than the minimum required to perform units conversion on the histogram and to put titles on a plot, this kept the contents relatively un-controversial. The GENIE-V3 workspace on the other hand is designed to support both one and two-dimensional histograms as well as several different models of neutron scattering data and even X-ray data. It is clear that without very careful design the workspace will be so general as to be no use to anyone!

5.3 Morphology of a workspace

The representation or model of an entity in GENIE consists of a structural component and a set of operations which act upon it. In the code of the program, these form a rigid set of associations, only changeable by re-compiling the code. The Number and Vector types in GCL correspond to these fixed associations built into the program. All models of entities other than these are implemented using the Workspace.

The GENIE-V3 workspace acts as a *flexible* vehicle for modelling neutron scattering entities. The contents of a workspace may be defined by the scientist as may the set of operations on that workspace. In Figure 3, all models shown above the Internal/External dashed line are implemented using GENIE-V3 workspaces. Those entities denoted in bold are already implemented as part of GENIE-V3 to give compatibility with GENIE-V2.

Using workspaces, GENIE-V3 provides a basic model of a histogram with its operations. The standard workspace also models a TOF spectrum to perform the same functions as GENIE-V2. To see how the workspace is used, three examples are looked at. The first is part of the standard GENIE-V3 GCL code to load a TOF spectrum with ISIS data from a raw file. The second shows a user defined procedure to load a workspace with an extra spectrum kept for normalizing the data. The third shows how the same user might re-define the "+" operation to ensure that the extra data is catered for whenever a workspace addition is carried out.

```
Example 1.

PROCEDURE S
PARAMETER Specno = INTEGER
! The procedure performs the function
!w1 = S(N)
! to load a time of flight spectrum.
...
! Example values
! selfile = "HET001345.RAW"
! specno = "10"
...
! Load TOF data into wtemp from selected raw
! file
Wtemp.X = Get:Data(selfile, "X-Array")
Wtemp.V = Get:Data(selfile, "Y-Array@"+specno)

! Generate statistical errors on count values.
Gen_errors(Wtemp.V)

! Get other TOF info
Wtemp.Two_theta = Get:Data(selfile, "2-THETA")
Wtemp.Delta = Get:Data(selfile, "DELTA")
Wtemp.L1 = Get:Data(selfile, "L1")
Wtemp.L2 = Get:Data(selfile, "L2")
Wtemp.X_units = TIME
Wtemp.Y_units = COUNTS_BY_W

!Return Wtemp as spectrum
Spectrum = Wtemp
ENDPROCEDURE Spectrum = WORKSPACE
```

In this example a simplified version of the GENIE-V3 "Get a spectrum" function is shown. The function takes a spectrum number as a parameter and loads a workspace with the data from the raw file. The parameters L1, L2 and Two_theta allow units conversions to be carried out on this workspace. There is no restriction on adding other Items to the workspace at any time.

In the second example. The user gets the normal data using the standard S function and then augments the workspace with the extra spectrum for normalization. The "+" function may now be augmented as shown in example 3. In this case it trivially normalizes the spectra after they have been added

```
Example 2.
PROCEDURE My_get
PARAMETER Specno = INTEGER
...
Wtemp = S(Specno)
Wtemp.XV = Get:Data("VAN.RAW", "X-Array")
Wtemp.YV = Get:Data("VAN.RAW", "Y-Array")
Spectrum = Wtemp
ENDPROCEDURE Spectrum = WORKSPACE
```

Note that if it exists, the function USER_PLUS is called automatically by the basic GENIE-V3 "+" routine, The parameters given to it are the two workspaces to be added with the result workspace ready to be ammended if necessary

```
Example 3.
PROCEDURE USER_PLUS
PARAMETER \                 ! "\" is continuation

V=WORKSPACE \              ! second workspace
Wtemp=WORKSPACE            ! result workspace

! Calculate an average normalization spectrum
! check the lengths just in case
IF Len(W.XV) = Len(V.XV)
      Wtemp.YV =  (W.YV + V.YV) / 2.0
ELSE
      Say  "Error in Add routine"
ENDIF

! and then apply it to the result.
Result.Y = Wtemp.Y / Wtemp.YV
ENDPROCEDURE Result = WORKSPACE
```

5.4 The Graphical view

Finally, an example of the Graphical view is shown. The graphical view is the total view the scientist has of the graphical representation of the GENIE internal models. This view does not just include the end result (a plot) but also includes the method of control. The example here shows a simple GENIE-V3 procedure using the graphics primitives to produce a plot of a workspace. The plot produced is shown in Figure 5.

```
Example 4.

PROCEDURE Display
PARAMETER Wout = WORKSPACE
! This procedure uses GENIE-V3 graphics
! primitives to create an example display
! of a workspace
!        Display S1
...

! Create Windows for large axes and draw plot
Win/Create/Unscaled 0.0 1.0 0.0 0.8 "GREY"
Win/Create/Scaled 0.2 0.8 0.1 0.6 "BACK"
Set/Label/X  Wout.Xtitle  Wout.Xunits
Set/Label/Y  Wout.Ytitle  Wout.Yunits
Draw/Axes/Auto  Wout.X  Wout.Y  "BLACK"
Draw/Hist  Wout.X  Wout.Y  "BLUE"  "SOLID"

! Now add small plot
Win/Create/Scaled 0.4 0.85 0.3 0.55 "YELLOW"
Draw/Axes 0.0 50.0 0.0 10E-9 "BLACK"
Draw/Box 0.0 0.0 1.0 0.6 "BLACK"
Draw/Box 0.0 0.6 1.0 0.748 "BLACK"
Draw/Box 0.6 0.5 0.8 0.6 "BLACK"
Draw/Axes/Auto  Wout.X  Wout.V  "BLACK"
Draw/Hist  Wout.X  Wout.Y  "RED"  "SOLID"

! Now add the title box
Win/Create/Unscaled 0.2 0.8 0.65 0.75 "BACK"
Draw/Text 0.15 0.72 Wout.Title1
Draw/Text 0.15 0.35 Wout.Title2

ENDPROCEDURE
```

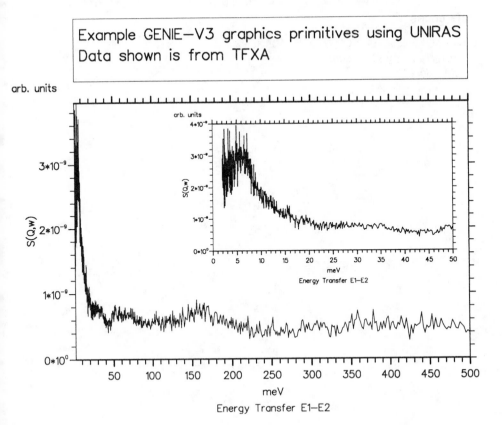

Figure 5.

References

David W I F et al. PUNCH GENIE Manual V2.3
   *Rutherford Appleton Laboratory report
   RAL-86-102.*

Inst. Phys. Conf. Ser. No 107: Chapter 1
Paper presented at Neutron Scatt. Data Anal. Conference, Rutherford Appleton, 1990

The visualization of neutron scattering data using UNIRAS

K.M.Crennell

Rutherford Appleton Laboratory, Chilton, Didcot,Oxon,UK

Abstract. NUVU is a data display program written in portable FORTRAN77 which allows users to see their two dimensional data in colour on any graphics device supported by the graphics package UNIRAS running on a VAX. UNIRAS has recently been purchased by the Combined Higher Education Software Team, (CHEST), and is now freely available to UK academic institutions. NUVU was written for use on the neutron scattering instruments at the spallation neutron source, ISIS, and should be in regular use in Spring 1990. Data can be seen in four time slices of the user's choice either by reading the data acquisition electronics during data collection or later by reading from a file. The graphics devices supported range from expensive workstations down to much cheaper IBM PC clones running the emulator program, EMUTEK, also available via CHEST. The main features of NUVU are described, including options to see plots of 1, 2 or 3 dimensional data, and use of image processing techniques to enhance the scientific understanding of the data.

1. INTRODUCTION - the need for 2D data display

The original neutron scattering instruments at ISIS, the pulsed neutron source at the Rutherford Appleton Laboratory, used linear detectors. Data analysis and display was performed using the program GENIE written by the instrument scientists and computer programmers.David *et al*(1986) GENIE graphics was based on the international standard Graphics Kernel System, ISO (1985), which defines low-level concepts for drawing lines and text in various styles and colours. GKS can display an image as a 'cell-array', but lacks any other routines for the display of 2 dimensional data and axes utilities.

Some of the newer instruments at ISIS contain area detectors, for example LOQ, a low-Q diffractometer used for measurements on macromolecules, biological and other large scale structures. The LOQ detector measures neutron counts using two arrays of wires forming a square matrix of side 128. The counts from each pair of wires are usually averaged together to make a data set of 64 by 64 numbers for each time interval measured from the time of arrival of the beam at the sample. Usually 100 intervals are recorded; they can be averaged together or used separately to investigate time-dependent scattering phenomena. LOQ users needed an easier, faster way of displaying their many data sets of the detector area than that obtainable using GKS for the low-level graphics. Other area detectors are being built, so it was clear that better graphical routines would have to be found or written to enable users to visualise their two dimensional data. Further details of ISIS instruments can be found in our annual report ISIS (1989) and in Finney (1989). Recently, Government funding has purchased the high level graphics package UNIRAS for use by all UK academics.

2. Description of UNIRAS

UNIRAS was originally written for seismology applications. It has many device drivers and high-level routines for the display of 1, 2 or 3D data including axes. Graphics can be saved in files using either the Computer Graphics Metafile, ISO(1987) format or its own proprietary format. Figure 1 shows the relationship between the various files used by UNIRAS which also has an implementation of GKS, so can be used immediately for applications already written to use GKS. However, since the GKS routines have been implemented in terms of UNIRAS own low- level routines, such plots are made more slowly then those written directly in the UNIRAS routines.

We are using Version 6.1e of UNIRAS which is supplied in 2 main parts, the subroutine library and the 'interactives', UNIMAP, the Picture manager, UPM, and UNIEDIT. A further interactive, UNIGRAPH, is promised in the next version, due in mid 1990. These 'interactives' give easier access to the routines from the subroutine library for users wanting to make simple 'one-off' plots. They present similar interfaces to the user, based on the concepts of windows as shown in Figure 1.

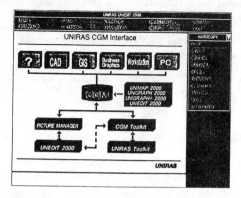

Figure 1. Example of the use of UNIEDIT to generate a flowchart describing the various files used by UNIRAS. The commands to generate hardcopy are seen on the right as they are on the display screen

Commands are either mouse or keyboard driven. Hardcopy is made from the current plot, but is not just a screen dump, the relevant parameters are saved to a file, a batch job submitted to convert those parameters to the format of the requested output device, and the plot printed automatically.

2.1. UNIEDIT

UNIEDIT is a drawing program making diagrams such as Figures 1 and 2. It resembles the drawing or 'painting' programs commonly found on popular microcomputers, such as MacDraw for the AppleMac. Figure 1 is supplied as one of the examples with UNIEDIT, it illustrates the use of the 'flowchart' icons to describe the various UNIRAS files and how they are related to each other. Figure 2 is an example of the usage of UNIEDIT to make typical diagrams of mineral structures. It was made by first drawing the two different octahedra which form the 'basic units' of the minerals, then using the 'replicate' feature to build up the rest of the diagram. UNIEDIT works with a conceptual plotting area of 20 meters, and has 'zooming' features to allow you to see the whole plot at a reduced size, so it can be used to design complete posters for Conference presentations. These can either be printed in small parts on A4 sized laser printers or all together as a single poster depending on the plotting devices available.

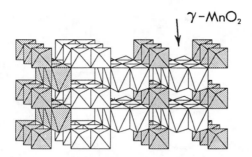

Figure 2. Two different arrangements of MnO_6, octahedra in two mineral forms for MnO_2, pyrolusite, (single edge sharing chains) and ramsdellite, (shaded, alternating double chains). The $\gamma - MnO_2$ is thought to be an irregular intergrowth of pyrolusite and ramsdellite.

Work is in progress at our laboratory to send such plots to the large plotter in another building normally used by electronic engineers for circuit designs.

3. Applications to ISIS instruments, NUVU

NUVU, (NeUtron Visualisation with Uniras) is a program which has been written in FORTRAN77 to use routines from the UNIRAS subroutine library to display output from the LOQ detector. It is intended for use during data collection on the instrument. Figures 3 and 4 show typical plots made with the 'image' output routine which displays individual pixels. They can be used to quickly see when any of the detector wires are malfunctioning, as in Figure 3.

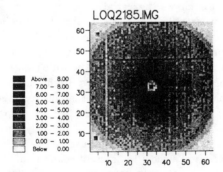

Figure 3. Plot showing two defective detector wires, one vertical on the left, the other horizontal just above the central area.

When data is being displayed from a file it may be better not to use the routine which displays the individual pixels, but to use the one which shows shaded filled contours which interpolates the individual points to give a smoother appearance to the boundaries between the shades. The LOQ detector elements are represented by a single data point which may not correspond to the centre of gravity of the neutron signal within the element, hence smoothing may give a better picture of the shape of the data, as seen in Figure 5.

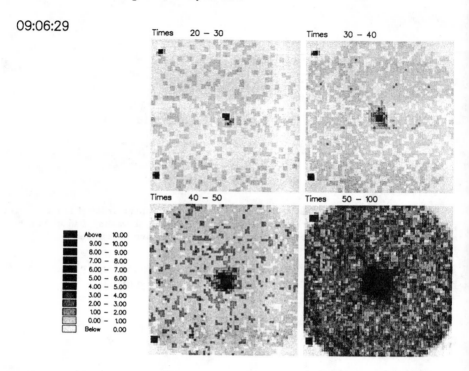

Figure 4. Plots showing neutron counts at the detector for four successive time periods measured in microseconds from the arrival of the ISIS beam pulse. The time when the data was read from the data acquisition electronics is seen at top left above the reference colour scale.

3.1. Shading scales and mapping

The scale to the left of the plots shows the variation of grey scale with neutron counts. This scale an be changed interactively to allow for the variation in counts seen when different samples are measured. Data can be displayed either before capture from the data acquisition electronics or later from a file. The colour scales needed may be different in the two cases. Screen displays cannot display the wide range of values seen in this data exactly; they vary between zero and thousands. High quality workstations typically display 256 different colours; it is doubtful whether the average human eye can distinguish between so many shades, so that there has to be a mapping between the number of colour scales available in the device, the 12 or so easily perceived by the average eye, and the counts to be displayed. The screen used on LOQ is an IBM PC clone with an EGA graphics board. It can show 16 different colours but there are often several hundred counts in the central area, so the user has to choose which parts of the detector to display in detail. Figure 4 taken during data collection shows the greater detail in the background area, with the central area all the same colour. To see detail in the peak at the same time as in the background needs a non-linear colour scale. A series of such scales is being developed, including the logarithmic scaling which has been used successfully in line contour plots made using GKS.

3.2. Use of false colour

These diagrams representing neutron counts have no natural colour, but the human eye is very good at seeing patterns in colour, so plotting them using 'false colour' can

3422.BIN

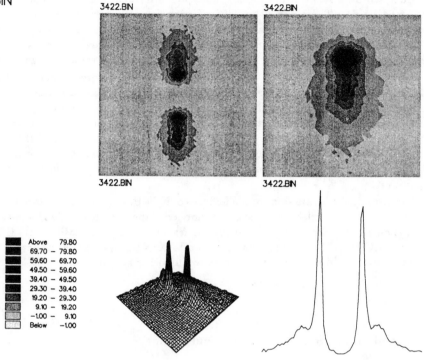

Figure 5. Various views of a two dimensional data set following correction for instrument parameters and geometry.(a) Top left: data as 2D shaded contours (b)top right:enlarged view of lower peak (c)Lower left: Data as a 3D surface, changed viewing angle and maximum values truncated in an attempt to show structure at the base of the peak. (d)Lower right: single line through the centre of the data in the Y direction showing peaks differ in height and have shoulders on them.

often bring out interesting features. UNIRAS has 6 standard colour scales used for different applications; LOQ data currently uses the first of these ranging from dark blue, through green, red, and yellow to white. User defined scales are possible, so we are experimenting with others which may be more appropriate for our application, particularly those already in common use, for example to display topographic information in Ordnance Survey maps, or those present in a rainbow.

3.3. Problems in displaying plots from analysed data.

The NUVU program can also display two dimensional data from stored files, these may be raw data files, as in Figure 4, or the result of data analysis after geometric and other instrument parameters have been taken into account. Figure 5 shows several views of an analysed data set. Plots may be seen filling the whole screen, or when more information is conveyed by having many plots on the screen the user can request as many as are needed. NUVU does not constrain the number of plots to any maximum, it attempts to scale the data to whatever is asked for. It may be anything from 1 plot on the screen to a matrix of 2 by 3 or even 10 by 10. The user should decide how much detail is needed and set the number of plots on the screen accordingly.

Analysed data sets can have many more points in them than the original raw data set of 64 by 64. There are 80 by 80 in Figure 5a, an analysed data set with some negative values. This is less than the resolution of the screens we are using, but when bigger data sets are generated they need to be scaled with care before display if vital features are not to be lost. NUVU has a 'zoom' facility, illustrated in Figure 5b.

NUVU allows interactive changing of the viewpoint of a three dimensional surface plot to find a better view, such as that in 5c, where the two peaks are seen to be clearly separated. However, some detail is still obscured in this view where the peaks appear to be the same height because they have been truncated in order to try to emphasize the features in the lower levels. The plot in Figure 5d is a histogram of the values in the data along the line through the centre. It clearly shows that the two peaks are not the same height and there are possible shoulders on their sides which may be physically significant.

Figures 3,4,5 illustrate some of the features of NUVU. Others allow addition, amendment or deletion of titles, sub-titles, axes and axis labelling. It is a flexible interactive tool, allowing the user to discover interesting features in the data easily. NUVU is written in FORTRAN77 and so far run only on a VAX/VMS system, but as only standard constructs are used it should be portable to any other system running UNIRAS.

4. Use of an image processing subroutine library

Two dimensional data similar to the data shown in figures 3 to 5 is used in many other scientific disciplines, including astronomy, medicine, and robotics. There is also a growing use for images in the fast developing field of desktop publishing. Although many people still insist on writing their own image analysis software, others are now turning to one of the image processing subroutine libraries, (Takamura et al 1983, Landy et al 1984, Crennell 1988) to save them time and effort. One of these, IPAL, the Image Processing Algorithms Library, was created by a collaborative project between NAg Ltd and the Rutherford Appleton Laboratory with funding from the 'Alvey Project' a five year research programme in Information Technology funded by the UK Department of Trade and Industry. This library has utilities for operations such as scaling, rotating, and combining images with common arithmetic operations. Other routines give statistics and histograms such as that in Figure 6d, the histogram of intensities present in the image of Figure 4. Data from neutron scattering is often noisy, and it may be useful to smooth using one of the routines from the 'Filters' section of IPAL. Figure 6b shows the data of Figure 3 following smoothing by a 3 by 3 median filter. Shapes can be clearly seen using thresholding, either interactively with the user specifying the number and value of levels, or using automatic routines which look for minima in the histogram as suitable values to threshold. I referred above to the problem of choosing the colour scales optimally to display data features. Contrast enhancement routines can give help with this problem. Figure 6c shows the data of figure 6b enhanced by the technique of histogram smoothing so that areas of uniform shading in figure 6b are now seen to have more structure. Other sections of IPAL include edge detection, transforms, and a selection of test patterns.

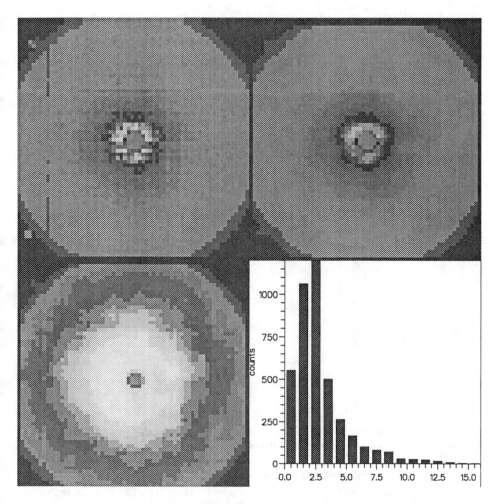

Figure 6. examples of use of image processing routines on the original data of Figure4, showing clockwise form top left, the original image, an image smoothed with a median filter, a grey level histogram of the image in top left and the result of contrast enhancement in lower left

5. Applications of UNIRAS during instrument installation and operation

Another program, VUMARI, has been written to summarise the state of the detectors for the instrument, MARI (ISIS 1989), now under construction. MARI, a multi-angle rotor instrument is being financed through an agreement with Japan. It is planned to have over 800 detectors when completed; some were installed during 1989, others during the shutdown in early 1990. VUMARI was written to give a graphical summary of the state of the detectors. Figure 7 shows a view of all the detectors, made from a run, (number 87) testing for noise in detectors. This is normally displayed on a colour screen, using 5 colours arranged in a 'traffic light' scale. Detectors not yet installed are shown as 'undefined', good ones, i.e. those with zero counts in this noise run, are shown in white. Ones with counts are divided into classes, settable by the user, shown in green, yellow and red. Red is reserved for the detectors with highest counts which need to be adjusted by the operations staff. They can look at the plot of Figure 7 and see immediately which part of the detector has most red lines and needs working on most urgently. Figure 7(right) is an example of the alternative plots used

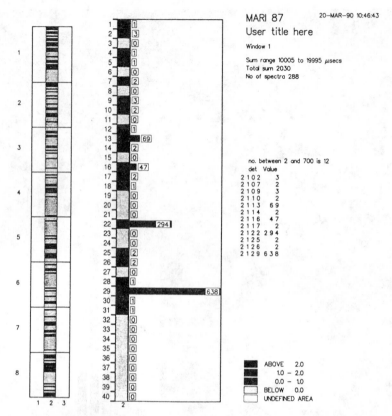

Figure 7. Examples of plots for MARI showing the state of the newly installed detectors, faulty ones have more than 10 counts. Left: complete instrument, right: a single window with detector list

to display subsets of the detectors, in this case the central strip, number 2, of the first window. The detector numbers are given on the right of the shaded representations of the detector counts. The counts are also shown as the histogram on the right of the shaded strip. The counts themselves are seen on the right of the histogram bars. These counts can be printed on request as a summary on the right of the page. Similar plots may be useful for displaying data from runs with samples in the detector, but these will need to use a different colour scale. We hope to explore this during the next ISIS operating cycle.

6. Conclusions and Future plans

The subroutine library of the UNIRAS package has been used to make useful plots of data from two ISIS instruments LOQ and MARI. We hope to have these in use for real user data during the ISIS cycle starting in April 1990. More work is needed to make a better user interface, to define a set of useful colour scales and to simplify the production of hardcopy. This may be done in a similar manner to that used by the UNIRAS 'interactives' which we have just installed, and which show great potential for improving our data display.

UNIRAS is a large complex package. Sections so far unexplored may be useful for more automatic plotting of mineral structures. The computer aided design primitives, including shaded spheres and cylinders, may become the basis for elegant molecular diagrams.

Acknowledgments

I am grateful for many useful discussions with ISIS instrument scientists, particularly, R.K.Heenan, A.D.Taylor and S.Hull. N.M.Hill, Central Computing Department, Rutherford Appleton Laboratory, has helped greatly in the mounting of the UNIRAS system on our computers and solving the display problems. I am grateful to NAg Ltd for use of a pre-release version of the IPAL library.

References

Crennell K M 1988 *The design and implementation of a portable image processing algorithms library* Image Processing 88 Proceedings of the Conference held in London. Blenheim Online ISBN 0 86353 157 1 p247-256

David W I F, Johnson M W, Knowles K J, Moreton-Smith C M, Crosbie G D, Campbell E P, Graham S P and Lyall J S 1986 *PUNCH GENIE Manual* RAL-86-102

Finney J L 1989 *ISIS a resource for neutron studies of Condensed Matter* Europhysics News **20**

ISIS annual report 1989 RAL-89-050 obtainable from the Laboratory library.

ISO IS 7943, 1985 *Information Processing Systems - Computer Graphics - Graphics Kernel System (GKS) functional description* ISO Geneva

ISO IS 8632 1987 *Information Processing Systems - Computer Graphics - Metafile for the storage and transfer of picture description information (CGM)* ISO Geneva

Landy M S, Cohen Y, Sperling G 1984 *HIPS: a UNIX-based Image Processing System* Computer Vision, Graphics and Image Processing **25** p331-347

Takamura H, Sakane S, Tomita F, Yokoya N, Kaneko M and Sakaue K 1983 *Design and Implementation of SPIDER - a Transportable Image processing Software Package* Computer Vision, Graphics and Image processing **23** p273-294

Extending the power of powder diffraction for structure determination

W.I.F. David,
Neutron Science Division, ISIS, Rutherford Appleton Laboratory, Chilton, Oxon, OX11 0QX,
U.K.

ABSTRACT

Using modern high resolution neutron and X-ray powder diffractometers, an impressive amount of structural information may be obtained that permits the refinement, using the Rietveld technique[1], of structures containing more than 100 positional parameters. Despite these technical advances and the renewed popularity of the technique, powder diffraction suffers from the inevitable loss of information imposed by the collapse of three dimensions of diffraction data on to the one dimension of a powder pattern. This can seriously frustrate the determination of unknown crystal structures. As a consequence of accidental or exact reflection overlap, only a relatively small number of reflections may have uniquely determined intensities. (For example, the cubic reflections (550), (710), and (543) exactly overlap.) In this paper, a maximum entropy algorithm (previously discussed theoretically[2]) is presented that evaluates, from first principles, the intensities of overlapping reflections in powder diffraction profiles. Indeed, good reconstruction may be retrieved when only ~20% of the reflections are uniquely determined. This precludes systems where the Patterson group is not the holosymmetric group (e.g. 4/m symmetry where hkl and khl reflections are overlapping but inequivalent.) Nevertheless, for the majority of crystal structures with unit cell volumes less than ~ 1000Å^3 this method should facilitate structure solution.

INTRODUCTION

Neutron powder diffraction is a powerful materials science technique that provides a detailed description of moderately complex crystal structures. This is nowhere more apparent than in the area of high temperature superconductivity which has brought neutron powder diffraction to the forefront of materials research over the past three years. However, the major milestone in neutron powder diffraction occurred twenty years ago with the development of the Rietveld profile refinement method, which has dominated the field of neutron powder diffraction. Recently, the advent over the past five years of high resolution instrumentation such as the high resolution powder diffractometer, HRPD, at ISIS has opened up new areas of study. In particular, the determination of completely unknown structures by powder diffraction is receiving growing attention and represents a substantial departure from structure refinement by the Rietveld method. In the Rietveld method it is necessary to have a reasonably good approximation to the crystal structure prior to least squares refinement. In ab initio structure determination no such assumptions about crystal structure can be made. Structure determination from powder diffraction data may be considered to proceed in five steps :

1. Derivation of the unit cell from measured Bragg peak positions using autoindexing methods.

2. Determination of the possible space group symmetry from examination of systematically absent reflections.

3. Least squares refinement of the intensities, I(h), and hence the structure factor magnitudes, $|F(h)|$, of all measured Bragg reflections. (N.B. I(h) α $|F(h)|^2$), .

4. Application of direct methods or Patterson methods of structure solution using the observed structure factor magnitudes.

5. Least squares refinement of the approximate crystal structure obtained from step 4.

The final two steps in structure solution are common to both single crystal and powder diffraction experiments. The penultimate step is oversimply described and may indeed be extremely convoluted since only the magnitude but not the phase of each structure factor is known. However, there is an additional complication in structure solution by powder diffraction techniques in obtaining structure factor magnitudes (step 3). The inevitable loss of information that results from the compression of three dimensions of crystallographic data on to the one dimension of a powder pattern can be a severe problem particularly with high symmetry systems. The purpose of the present work is to provide an optimal method for overcoming this problem. Relationships exist between different   structure factor magnitudes because of properties such as atomicity and, in the case of X-ray scattering, positivity of scattering density. It is thus to be expected that an algorithm that harnesses all the structure factors simultanelously and in a rigorous manner with respect to loss of information associated with peak overlap will provide information about the individual intensities of overlapping reflections. A logical method is to proceed via the Patterson function, the Fourier transform of the squares of the structure factor magnitudes, $|F(h)|^2$. The maximum entropy algorithm discussed in the following section fulfils all the above conditions. Experience indicates that good reconstruction may be retrieved when only ~20% of the reflections are uniquely determined and should be applicable to the majority of crystal structures with unit cell volumes less than ~ 1000Å^3 .

THE MAXIMUM ENTROPY ALGORITHM

Crystallography is rooted in the mathematics of the Fourier transform. The structure factor, F(h) (in general a complex quantity), that is associated with reflection h, is the Fourier coefficient of the atomic structure $\rho(x)$

$$F(\mathbf{h}) = \int_V \varrho(\mathbf{x}) \exp(2\pi i \mathbf{h}.\mathbf{x}) d^3 x$$

In a crystallography experiment the measured quantity is $|F(h)|^2 = F(h)F^*(h)$ and the phase information is lost. $|F(h)|^2$ is the Fourier component of the Patterson function, $P(u)$, the autocorrelation function of the scattering density (or, alternatively, the interatomic vector map of the crystal structure).

$$|F(\mathbf{h})|^2 = \frac{1}{2} \int_V P(\mathbf{u}) \cos(2\pi \mathbf{h}.\mathbf{u}) d^3 u$$

In a powder diffraction experiment many reflections may overlap and, in general, measured observations are of the form

$$G_i = \sum_{n=1}^{N_i} J(\mathbf{h}_n) |F(\mathbf{h}_n)|^2$$

$$= \int_V P(\mathbf{u}) \left[\sum_{n=1}^{N} \left(\sum_{j=1}^{J(\mathbf{h}_n)} \cos(2\pi \mathbf{h}_{n_j}) \right) \right] d^3u \qquad 1$$

Although the observations are no longer Fourier coefficients of any meaningful function, equation 1 is nevertheless linear and, if the Patterson map is divided into N_{PIX} pixels may be expressed in the form

$$G_i = \sum_{k=1}^{N_{PIX}} w_{ik} P_k \qquad (i = 1,...,N_{OBS}) \qquad 2$$

The coefficients w_{ik} correspond to the quantity within square brackets in equation 1. The method proposed in this letter is to maximise the 'Patterson entropy'

$$S = - \sum_{k=1}^{N_{PIX}} P_k \ln(P_k/Q_k) \qquad 3$$

(Q_k is the a priori estimate, assumed flat, of the Patterson map) while maintaining faithful agreement with the observed data D_i (standard deviation σ_i) through the chi–squared constraint

$$\sum_{i=1}^{N_{OBS}} \frac{1}{\sigma_i^2} (D_i - G_i)^2 = N_{OBS}$$

The maximum entropy technique is a powerful method of image reconstruction that has been successfully applied to a number of different areas of research ranging from radio astronomy[3] to crystallography. The application of maximum entropy to crystallography has been discussed by several authors (see, for example, the work by Livesey and Skilling[4], and Bricogne[5]. The above formulation of the maximum entropy algorithm (equation 3) implicitly assumes that the scattering density is everywhere positive. Although this is true for X–ray diffraction analysis, negative Patterson densities may result from the analysis of neutron powder diffraction data. In such cases the Patterson map is divided into two arrays with intrinsically positive and negative contributions. Equations 2 and 3 are then recast as

$$G_i = \sum_{k=1}^{N_{PIX}} w_{ik} (P_k^+ - P_k^-) \qquad (i = 1,...,N_{OBS}) \qquad 2a$$

$$S = - \sum_{k=1}^{N_{PIX}} \left[P_k^+ \ln(P_k^+/Q_k^+) + P_k^- \ln(P_k^-/Q_k^-) \right] \qquad 3a$$

The final step in the full three-dimensional reconstruction of the diffraction data is achieved by calculating individual Fourier components of the maximum entropy Patterson and thus separating the intensities of overlapping reflections; i.e.

$$|F(\mathbf{h}_n)|^2 = \frac{1}{2} \int_V P_{ME}(\mathbf{u}) \cos(2\pi \mathbf{h}_n.\mathbf{u}) d^3u \qquad 4$$

EXAMPLE: RUTILE

As an example of the method, an analysis of medium resolution neutron powder diffraction data is presented. The material, rutile (TiO_2), was selected for two reasons: firstly, the relatively high tetragonal symmetry (P4$_2$/mnm : a = 4.5929Å, c = 2.9581Å) of rutile and medium resolution of the dataset collected on the POLARIS[6] diffractometer at ISIS ensures a reasonable degree of reflection overlap; and secondly, the presence of positive (O) and negative (Ti) scattering centres tests the generality of the method.

Diffraction data from a 5cm^3 sample of rutile were obtained from the POLARIS diffractometer at ISIS in 2 hours. After normalisation, the profile was fitted refined using the technique devised by Pawley[7]; in addition to Bragg peak position (determined by unit cell size) and peak width variables, the intensities of individual Bragg reflections were refined as separate variables with the proviso that reflections that were close together (typically less than 10% of the full–width at half maximum) were constrained to have equal structure factor magnitudes. (This is the cor-rect *a priori* Bayesian assumption.) 177 reflections were recorded in 130 clumps; 97 reflections were not overlapping. Comparison of the observed and calculated diffraction patterns (Figure 1) indicates an excellent agreement (χ^2 = 1.37). (From the density of the tick marks, which indi-cate the Bragg reflection positions, it is clear that there is considerable reflection overlap particu-larly at short d spacings. By eye there appears to be only around 65 peaks in the full pattern!)

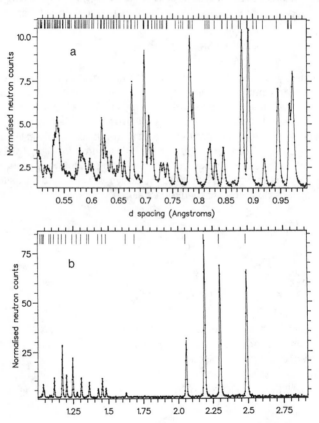

Figure 1
The observed (dots) and calculated (line) profiles for TiO$_2$ obtained using the POLARIS powder diffractometer at ISIS. The d-spacings range (a) from 0.5Å to 1Å and (b) from 1Å to 2.9Å. The tags mark the positions of Bragg reflections and clearly indicate the severe peak overlap particu-larly at short d-spacings.

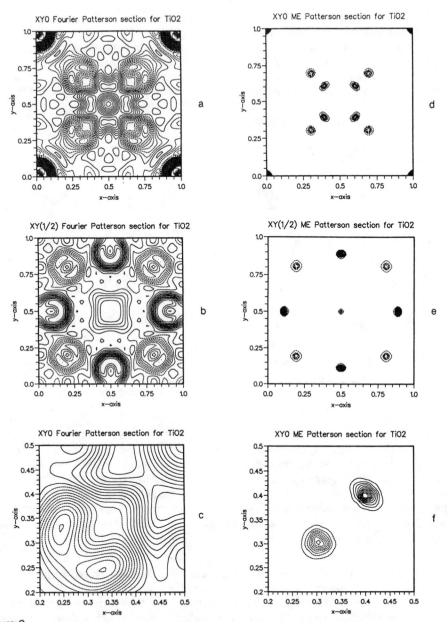

Figure 2

Figures 2a–c: Sections of the Patterson map synthesised by Fourier transformation of the squares of the structure factor magnitudes obtained from the Pawley-type profile refinement of TiO_2 and subsequent averaging of a sphere of diameter 0.5Å. The broad structural features indicate approximate vector positions but are severely distorted by misapportioning of intensities and truncation errors. The first two maps range over the full xy plane at (a) $z = 0$ and (b) $z = \frac{1}{2}$. Figure 2c is a magnification of Figure 2a with $0.2 < x < 0.5$ and $0.2 < y < 0.5$. Figures 2d–g: Sections of the Patterson map obtained by maximum entropy reconstruction from the observed clumps of Bragg intensities. The features are extremely sharp and are correctly located at interatomic vector positions. The map ranges in Figures 2d–g correspond to those of Figures 2a–c.

Figures 2a–c show sections of the Patterson map obtained by direct Fourier transformation of the equipartioned squares of the structure factor magnitudes obtained from the Pawley refinement. Broad features dominate the maps with clear evidence of oscillations resulting from Fourier truncation errors and splitting of interatomic vector peaks caused by misapportioning of Bragg intensities. It is not possible to locate the interatomic vectors (listed in Table 1) with any degree of confidence. Using a uniform prior distribution, a maximum Patterson map consistent with the observed data was obtained, by iteration, in 64 cycles. The evolution of the entropy and chi-squared is presented in Figure 3.

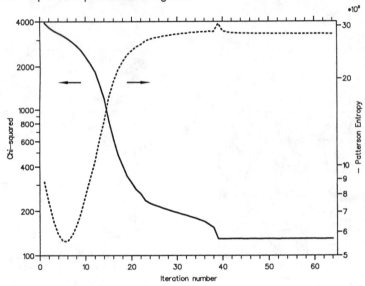

Figure 3
Evolution of the entropy and chi-squared goodness of fit for the construction of the maximum entropy Patterson map. Both quanitities are plotted on a logarithmic scale. (The negative value of the Patterson entropy is plotted as it is an intrinsically negative property.) Note that the entropy decreases initially as the Patterson map changes from the flat uniform prior distribution to one consistent with the observed data ($\chi^2 = 130$) after 39 iterations. Once agreement with the observed diffraction intensities has been obtained the entropy increases to its maximum value consistent with the χ^2 constraint.

The Patterson maps (Figures 2d–g) obtained using the maximum entropy algorithm contrast markedly with the Patterson maps derived by direct Fourier transformation. Sharp positive and negative features are located close to the the correct interatomic vector positions. Indeed, it is clear from examination of Table 1 that the typical deviation between observed and correct vector positions is ~0.0015a = 0.007Å, a full order of magnitude less than the pixel resolution (a/60 ≃ 0.08Å) and almost two orders of magnitude less than the minimum d-spacing measured (0.487Å). The oxygen fractional coordinate calculated from the maximum entropy Patterson map is 0.3047 essentially identical to the correct value of 0.3048. Moreover, the intensities of these interatomic vectors are in good agreement with expected values (Table 1) and there is even some evidence of the anisotropy of the thermal motion of the oxygen atoms in the positive feature at (0.3949,0.3949,0.0000) (see Figure 2g). The algorithm has produced an optimally sharpened Patterson map and thus opens up a productive Patterson methods route for the solution of crystal structures.

Table 1

Comparison of correct and observed (ex maximum entropy Patterson map) interatomic vector positions and intensities for TiO_2

ME Map	x	y	z	Δ	ME intensity	True intensity
Origin Peak	0.0000	0.0000	0.0000	0.0000	1.671	1.147
Ti – O	0.3041	0.3041	0.0000	0.0010	-0.387	-0.399
O – O	0.3949	0.3949	0.0000	0.0064	0.325	0.337
Ti – Ti	0.5000	0.5000	0.5000	0.0000	0.166	0.118
Ti – O	0.1942	0.1942	0.5000	0.0014	-0.358	-0.399
O – O	0.5000	0.1110	0.5000	0.0014	0.667	0.673

Δ = fractional difference between ME and true interatomic vectors

Evaluation of the individual structure factor squared magnitudes was performed by back trans-formation of the maximum entropy Patterson map (equation 4). A comparison of the following R-factors shows that there is a clear superiority in the $|F(\mathbf{h})|^2$ derived from the maximum entropy algorithm over the *a priori* equipartioned values.

$$R_{EQ} = \left[\sum^{N_F} w(\mathbf{h}) \left(F(\mathbf{h})_{EQ} - F(\mathbf{h}) \right)^2 / \sum^{N_F} w(\mathbf{h}) F(\mathbf{h})^2 \right]^{\frac{1}{2}} = 34.0\%$$

$$R_{ME} = \left[\sum^{N_F} w(\mathbf{h}) \left(F(\mathbf{h})_{ME} - F(\mathbf{h}) \right)^2 / \sum^{N_F} w(\mathbf{h}) F(\mathbf{h})^2 \right]^{\frac{1}{2}} = 14.1\%$$

$$R_{EX} = \left[N_F / \sum^{N_F} w(\mathbf{h}) F(\mathbf{h})^2 \right]^{\frac{1}{2}} = 6.5\%$$

R_{EQ} and R_{ME} are the R-factors for the $|F(\mathbf{h})|$ obtained from Pawley profile refinement and maximum entropy algorithms respectively and R_{EX} is the expected R-factor based on statistical errors. N_F is the number of different $|F(\mathbf{h})|$ and w($\mathbf{h}$) is the inverse square of the estimated stan-dard deviation of $|F(\mathbf{h})|$. Moreover, for all clumps of reflections the method correctly discrimates between strong and weak intensities albeit usually with a slightly reduced contrast than for the true values. Indeed, the $|F(\mathbf{h})|^2$ derived from the maximum entropy algorithm for intensity clumps containing up to five reflections agree well with true values even at short d-spacings close to the limits of the diffraction data. A further indication of the success with which the maxi-mum entropy algorithm has optimally combined the diffraction intensity information is provided by consideration of the sign of $|F(\mathbf{h})|^2$. Although the least-squares profile refinement has yielded negative values for some $|F(\mathbf{h})|^2$, all 177 $|F(\mathbf{h})|^2$ obtained by back-transformation of the maximum entropy Patterson map are positive quantities. In general, therefore, the $|F(\mathbf{h})|^2$ obtained by the maximum entropy method can be used to increase the potential for success of a direct methods approach to structure solution from powder data.

CONCLUSION

Finally, it is worth considering two broader implications of the maximum entropy formalism presented in this paper that may have important ramifications for *ab-initio* structure solution. The Patterson maps produced may be regarded as being optimally sharpened and thus be the closest approach to a true interatomic vector map. Although computationally expensive, this has broad implications for Patterson methods and will assist in the current revival[8] of the technique as an important alternative to direct methods of structure solution[9,10]. Furthermore, the formalism harnesses the information content in neutron diffraction experiments of atoms possessing negative scattering factors. Many aspects of direct methods of structure solution from X-ray data rely implicitly upon the inherent positivity of the electron scattering density. This poses concerns for neutron diffraction data where the scattering density may be both positive and negative and, indeed, weakens the power of direct methods in such situations. In contrast to this in the present work, negative scattering density is a bonus. Consider, for example, a molecule occupying the asymmetric unit of a unit cell and containing a single atom with a negative scattering length. The negative peaks in the maximu6m entropy Patterson map only correspond to vectors between the negative scatterer and all the other atoms. Such a situation bears a close resemblance to isomorphous replacement[11,12] and anomalous dispersion[12] methods. Examination of the negative Patterson features uniquely defines the location and orientation of the molecule with respect to the negatively scattering atom.

REFERENCES

1. Rietveld, H.M. *J.Appl. Crystallogr.* **2**, 65–71 (1969)
2. David, W.I.F. *J. Appl. Crystallogr.* **20**, 316–319 (1987)
3. Gull, S.F. & Daniell, G.J. *Nature* **272**, 686–690 (1978)
4. Livesey, A.K. & Skilling, J. *Acta Cryst.* **A41**, 113–122 (1985)
5. Bricogne, G. *Acta Cryst.* **A40**, 410–445 (1984)
6. Hull, S. & Mayers, J. *Rutherford Appleton Laboratory Report* RAL–89–118 (1989)
7. Pawley, G.S. *J. Appl. Crystallogr.* **14**, 347–361 (1981)
8. Egert, E. *Acta Cryst.* **A39**, 936–940 (1983)
9. Woolfson, M.M. *Acta Cryst.* **A43**, 593–612 (1987)
10. Ladd, M.F.C & Palmer, R.A. (editors) *Theory and Practice of Direct Methods in Crystallography* (Plenum, New York & London, 1980)
11. Rossmann, M.G., Arnold, E. & Vriend G. *Acta Cryst.* **A42**, 325–334 (1986)
12. Karle, J. *Acta Cryst.* **A45**, 765–781

Inst. Phys. Conf. Ser. No 107: Chapter 2
Paper presented at Neutron Scatt. Data Anal. Conference, Rutherford Appleton, 1990

101

Maximum entropy reconstruction of spin density maps in crystals from polarized neutron diffraction data

R. J. Papoular and B. Gillon

Léon Brillouin Laboratory, CEA - CNRS, CEN/Saclay, 91191 - Gif/Yvette Cédex, France.

ABSTRACT: The Maximum Entropy Method (MaxEnt) has been applied to the determination of spin density maps in centrosymmetric single crystals from polarised neutron diffraction data. By comparison with the Standard Inverse Fourier method, MaxEnt yields much cleaner maps by at least one order of magnitude. One can now "see" the magnetisation in a crystal, which was not possible using the standard procedure, due to very severe truncation effects. Potential benefits for studies involving high-Tc superconductors or molecular complex compounds are shown to be enormous.

1.INTRODUCTION

Polarised Neutron Diffraction (P.N.D.) aims at retrieving the magnetisation density or the spin density of given single crystals from limited and noisy sets of Fourier components (Brown *et al* 1980, Gillon and Schweizer 1989). Truncation effects are very severe, forbidding the use of a straightforward inverse Fourier transform in most of the cases. Nonetheless, inverse Fourier transforming remains one of the two approaches in current use. Smoothing or apodization has been used for some time in conjunction with the latter (Shull and Mook 1966), but it results in a loss of spatial resolution, which is highly undesirable in fields of current interest such as molecular magnetism or high-Tc superconductivity. In the aforementioned cases, the expected features of the sought magnetisation density, crucial to check current theories, should occur as small peaks which would be wiped out by smoothing. A second shortcoming of a straight Fourier inversion is that the experimental error bars cannot be taken into account. The second approach in current use is model fitting. Here again, severely limited data sets restrict the number of possible parameters which in turn constrain the results, which are thus biased. Moreover, the resulting best parameters (most often in the least-squares sense) are strongly correlated and thus unreliable to some extent. Another procedure is thus called for. In order to avoid the drawbacks mentioned above, the following properties are required: i) Nothing should be assumed about what is not measured, if not part of our prior knowledge. Truncation results from forcing all the unmeasured Fourier components to nil. ii) Experimental error bars should be taken into account. iii) The procedure should be model-free. iv) It should be able to incorporate our prior knowledge regarding the magnetisation density. v) The solution should be unique.

Such a procedure exists and has been increasingly developed up to now, as demonstrated by Skilling and Gull's recent work (Skilling and Gull 1989): It is Maximum Entropy (MaxEnt), sometimes also known as the Maximum Entropy Method (M.E.M.) or the Maximum Entropy Principle (M.E.P.). Maximum Entropy, already introduced at the first Workshop on Neutron Scattering Data Analysis (Johnson 1986), has been claiming more and more successes in such fields as Astronomy (Frieden and Swindell 1976, Gull and Daniell 1978, Narayan and Nityananda 1986), N.M.R. (Laue *et al* 1985,1986) and more generally Image Reconstruction since the work of Burg (1967).

Maximum Entropy addresses the following problem: what is the "smoothest", "least informative", positive distribution (the Image) compatible with a given set of constraints (the Data: e.g. experimental data, with suitable error bars) and our prior knowledge (e.g. symmetry requirements) about this distribution? A crucial property is the uniqueness of the solution, when the constraints are linear.

The Polarised Neutron Diffraction problem has this latter property, and it is the purpose of this paper to report on the first successful applications of MaxEnt to this problem.

To conclude this introduction, the reader is reminded that MaxEnt has already been used in conjunction with neutron scattering data in the following cases: small-angle scattering, liquid and amorphous material diffraction, quasielastic scattering, inelastic scattering, and even neutron spectrometer design.

2. THE POLARISED NEUTRON DIFFRACTION PROBLEM

2.1 The experiment

Its purpose is to collect data sets of Fourier components $F_m(\vec{K})$'s of the sought magnetization density, together with the related experimental error bars $\sigma(\vec{K})$'s, as a function of the scattering vector $\vec{K}$.

The experiment consists in impinging a monochromatic beam of initially polarised neutrons on a single crystal subject to a large vertical magnetic field $\vec{H}$. The crystal is set so that a crystallographic direction $\vec{n} = (uvw)$ is vertical. Accordingly, the horizontal scattering plane P is a plane from the reciprocal lattice. The scattered neutrons are collected at the detector for those discrete directions corresponding to the Bragg reflections (hkl) associated to scattering vectors $\vec{K}$. These Bragg reflections result from the long range spatial ordering within the single crystal. The applied magnetic field, inducing a magnetisation in the sample, makes the scattering dependent upon the initial polarisation state ($\uparrow$) or ($\downarrow$) of the incoming neutrons. For each Bragg reflection the two intensities $I_\uparrow$ and $I_\downarrow$ are collected, yielding the flipping ratio $\mathcal{R} = I_\uparrow / I_\downarrow$, together with its experimental error bar.

In the case of a centrosymmetrical single crystal, the 3-D Fourier components of the magnetisation density $m(\vec{r})$ (also called the Magnetic Structure Factors) sought for are real and can be deduced most often unambiguously from the related flipping ratios. The associated error bars are obtained using the law of propagation of errors, assuming gaussian errors throughout. We shall consider only centrosymmetrical crystals in what follows.

Quite a similar problem is the retrieval of the Patterson function in direct space, from the intensities of magnetic or nuclear Bragg peaks (instead of magnetic or nuclear structure factors). Either the 3-D Patterson function or 2-D projections can be sought, exactly as in the Polarised Neutron Diffraction case. Moreover, the Patterson function is always centrosymmetric. The mathematics involved are rigourously identical.

The Magnetic Structure Factors obtained from Bragg reflections belonging to the scattering plane P can also be considered as 2-D Fourier components of the projected magnetisation density $p(\vec{r})$ along the crystallographic direction (uvw). For practical and computational reasons, the present paper will deal only with the 2-D reconstruction of the projected magnetisation density $p(\vec{r})$. The mathematics for the 3-D case are quite similar to those for the 2-D case, but the amount of required experimental data and the size of the computations are substantially higher: many more Bragg reflections must be collected and the size of the image sought, now three-dimensional, is about a hundred times larger.

The limitations of the reciprocal space which is probed in a P.N.D. experiment, and hence the restricted data sets, result mainly from three causes: i) deriving the Magnetic Structure Factor from the flipping ratio for a given scattering vector $\vec{K}$ requires a measurable Nuclear Structure Factor $F_N(\vec{K})$ for the same $\vec{K}$ value. When this is not the case, the Magnetic Structure Factor cannot be obtained. ii) The experiment is carried out at a fixed value of the incident neutron wavelength λ, limiting the $\vec{K}$-space which is probed. iii) Lastly, the duration of the experiment is finite, which results in a limited counting statistics $\sigma(\vec{K})$.

2.2 Notations and basic equations

Let the crystal be described by a unit cell $(\vec{a}, \vec{b}, \vec{c})$ in direct space and $(\vec{a^*}, \vec{b^*}, \vec{c^*})$ be the related basis in reciprocal space. The crystallographic direction $\vec{n} = (u, v, w)$ is set perpendicular to the horizontal scattering plane P. Let $F_m(\vec{K}) = F_m(hk\ell)$ be the magnetic structure factor corresponding to the Bragg reflection $(hk\ell)$ associated with the scattering vector $\vec{K}$, and let $\sigma(\vec{K})$ be the related error bar. Let $m(\vec{r})$ and $p(\vec{r})$ be respectively the 3-D magnetisation density and the 2-D projected magnetisation density along the $\vec{n}$ direction. A second basis in direct space $(\vec{A}, \vec{B}, \vec{C})$ is required, so that $\vec{A}$ and $\vec{B}$ lie in the horizontal plane P and so that $\vec{C}$ is parallel to $\vec{n}$. $C = |\vec{C}|$ is equal to the shortest distance between lattice points along $\vec{n}$. Let V be the volume of the unit cell. The projected surface S of the unit cell onto the scattering plane P is then equal to V/C. It should be noted that the volume corresponding to the second basis $(\vec{A}, \vec{B}, \vec{C})$ is not equal to V in general, but to $M \cdot V$, where M is an integer. Lastly, let any point $\vec{r}$ in direct space be described respectively by the coordinates (x, y, z) and (X, Y, Z) in the two bases in direct space mentioned above.

Now, using the following relations:

$$\vec{r} = x\vec{a} + y\vec{b} + z\vec{c} = X\vec{A} + Y\vec{B} + Z\vec{C}$$
$$\vec{n} = \vec{C} = u\vec{a} + v\vec{b} + w\vec{c}$$
$$\vec{K} = h\vec{a}^* + k\vec{b}^* + \ell\vec{c}^*$$
$$\vec{K} \cdot \vec{r} = 2\pi(hx + ky + lz)$$

one obtains:

$$F_m(\vec{K}) = \int\int\int_\varphi d\vec{r}\ \exp\{i\vec{K}\cdot\vec{r}\}\ m(\vec{r}) \tag{1a}$$

$$= V \int_0^1 \int_0^1 \int_0^1 dxdydz\ \exp\{i\vec{K}\cdot\vec{r}\}\ m(\vec{r}) \tag{1b}$$

$$= S \int_0^1 \int_0^1 dXdY\ \exp\{i\vec{K}\cdot\vec{r}\}\ p(\vec{r}) \tag{2}$$

where :

$$p(\vec{r}) = C \int_0^1 dZ\ m(\vec{r}) \tag{3}$$

$$m(\vec{r}) = 1/V \sum_{\vec{K}} \exp\{-i\vec{K}\cdot\vec{r}\}\ F_m(\vec{K}) \tag{4}$$

$$p(\vec{r}) = 1/S \sum_{\vec{K}\in P} \exp\{-i\vec{K}\cdot\vec{r}\}\ F_m(\vec{K}) \tag{5}$$

2.3 Discretisation and reduction to the canonical form $\mathbf{D} = \mathbf{A} \cdot \mathbf{I} + \mathbf{n}$ †

As already stated above, we shall be concerned in what follows with the retrieval of the 2-D projected magnetisation density $p(\vec{r})$, rather than with the more general 3-D magnetisation density $m(\vec{r})$. We do not seek an analytical function $p(\vec{r})$, but a discrete set of real numbers $I_j = p(\vec{r}_j)$, where $1 \leq j \leq M$. The $\vec{r}_j$'s are defined as follows. Let us first recall that $p(\vec{r})$ depends only on the X and Y coordinates. Those coordinates map the scattering plane P, and the relevant directions are $\vec{A}$ and $\vec{B}$. Only that portion of the plane P corresponding to both $0 \leq X \leq 1$ and $0 \leq Y \leq 1$ is considered. The corresponding area is equal to M times that of the projected unit cell, which is equal to S. The two related intervals are binned into M_X and M_Y bins of sizes ΔX and ΔY respectively, yielding $M = M_X \cdot M_Y$ elementary cells or pixels. A given pixel can be described either by two integers j_X $(1 \leq j_X \leq M_X)$ and j_Y $(1 \leq j_Y \leq M_Y)$, or by a single one $j = (j_Y - 1) \cdot M_X + j_X$, with $1 \leq j \leq M_X \cdot M_Y$. Then, $\vec{r}_j$ corresponds to the center of pixel j.

---

† Data = Apparatus · Image + noise

In the simpler case when all the sought I_j's are strictly positive real numbers, the related set of $M_X \cdot M_Y$ I_j values is called an Image. The size of the latter is then $M_X \cdot M_Y$. In the more general (and intricate) case when each of the I_j's can be of either sign, each of the I_j is further defined as the difference of two strictly positive numbers $I_{j\uparrow}$ ($= p_\uparrow(\vec{r}_j)$) and $I_{j\downarrow}$ ($= p_\downarrow(\vec{r}_j)$) so that $I_j = I_{j\uparrow} - I_{j\downarrow}$. The combination of these two sets ($I_{j\uparrow}$'s and $I_{j\downarrow}$'s) constitutes the Image, the size of which is $2 \cdot M_X \cdot M_Y$. A single index i can be used, for instance defined by $i(j\uparrow) = 2j$ and $i(j\downarrow) = 2j + 1$.

Given any function $p(\vec{r})$, the calculated structure factors $F_m^c(\vec{K})$ can be computed for any set of $\vec{K}$ vectors belonging to $\mathcal{P}$, using formula (2). After discretisation, one may write in a compact matrix form:

$$\mathbf{F}_m^c(\vec{K}_k) = \mathbf{A}(k,j) \cdot \mathbf{p}(\vec{r}_j) \tag{6a}$$

where:

$$A(k,j) = S \cdot \Delta X \Delta Y \cdot \exp\{i\vec{K}_k \cdot \vec{r}_j\} \tag{6b}$$

Let us now turn to a given data set, corresponding to a set of $\vec{K}$'s belonging to the scattering plane $\mathcal{P}$. In practice, some of these $\vec{K}$'s are related through symmetry operations, resulting in a non independent set of $F_m(\vec{K})$'s. This apparent redundancy is used to check the consistency of the experiment, the occurence and evaluation of extinction effects and so on. Eventually, an independent set of N scattering vectors $\vec{K}$'s is derived, together with new sets of $F_m(\vec{K})$'s and $\sigma(\vec{K})$'s corresponding to better statistics. This is due to the fact that $F_m(\vec{K})$'s corresponding to symmetry related $\vec{K}$'s are essentially the same within a known phase factor. In the centrosymmetrical case considered here, this phase factor is equal to either 1 or -1 . How this new set is derived from the overall measured one is discussed in the next section (2.4). Let us consider the k-*th* independent $\vec{K}_k$ scattering vector and the related $D_k = F_m(\vec{K}_k)$ and $\sigma(\vec{K}_k)$ values deduced from experimental data. That a given projected spin density map $p(\vec{r})$ agrees with the experimental data within error bars is expressed mathematically by:

$$F_m(\vec{K}_k) = F_m^c(\vec{K}_k) + n_k \tag{7}$$

where n_k is a noise term of the order of $\sigma(\vec{K}_k)$. Equation (7) amounts to the announced matrix form with $\mathbf{F_m} = \mathbf{D}$ and $\mathbf{p} = \mathbf{I}$.

2.4 Reduction from a non-independent to an independent set of $F_m(\vec{K})$'s

The problem is the following: N equivalent magnetic reflections are measured, which pertain, within a known phase factor, to the same physical quantity — the magnetic structure factor of **one** of these equivalent reflections. How can we combine these different measurements of essentially the same quantity, to get the 'best' estimate F for this magnetic structure factor, together with a better accuracy σ_F than each of the independent measurements? One possible answer is to use Bayes' Rule to invert conditional probabilities. Hence, we seek the following probability law:

$$Prob(F|F_1,\sigma_1; F_2,\sigma_2; F_3,\sigma_3; \cdots) \propto \tag{8a}$$
$$Prob(F_1|F; F_2,\sigma_2; F_3,\sigma_3; \cdots) \cdot Prob(F_2|F; F_3,\sigma_3; \cdots) \cdots Prob(F_N|F) \cdot Prob(F)$$

In the (generally assumed) case of independent measurements, the formula above simplies further to:

$$Prob(F|F_1, \sigma_1; F_2, \sigma_2; \cdots) \propto \tag{8b}$$
$$Prob(F_1|F) \cdot Prob(F_2|F) \cdots Prob(F_N|F) \cdot Prob(F)$$

Now, $Prob(F_1|F) = \frac{1}{2\pi\sqrt{\sigma_1}} \exp -\frac{(F-F_1)^2}{2\sigma_1{}^2}$ and so on. Prob(F) is the a priori probability, before the measurements were carried out. Due to our complete ignorance prior to acquiring data, it is taken as a constant within the range over which the likelihood $Prob(F_1|F) \cdot Prob(F_2|F) \cdots Prob(F_N|F)$ is significantly different from 0. It follows immediately that the sought probability law for F is normal, is centered at $\overline{F}$ and has a standard deviation σ_F, such that:

$$\overline{F} = \frac{\sum\limits_{k=1}^{N} \left[\frac{F_k}{\sigma_k{}^2}\right]}{\sum\limits_{k=1}^{N} \left[\frac{1}{\sigma_k{}^2}\right]} \quad \text{and} \quad \frac{1}{\sigma_F{}^2} = \sum\limits_{k=1}^{N} \left[\frac{1}{\sigma_k{}^2}\right] \tag{9}$$

2.5 Incorporation of spatial symmetry requirements

The magnetic structure factors $F_m(\vec{K})$ are collected at scattering vectors $\vec{K}$ corresponding to nuclear Bragg peaks. Hence, the magnetisation density $p(\vec{r})$ to be reconstructed † must retain the nuclear structure symmetry of the single crystal under investigation, which is described by a factor space group $\mathcal{G}$. Since our focus here is on the projected magnetisation density along a given direction of the direct lattice, it should be noted that not all symmetry operations belonging to $\mathcal{G}$ are compatible with the projection. A compatible symmetry operation transforms any scattering vector $\vec{K}$ from the scattering plane $\mathcal{P}$ into another scattering vector belonging to that plane. These compatible symmetry operations constitute a sub-group $\mathcal{G}'$ of $\mathcal{G}$. Let h' be the order of $\mathcal{G}'$. Moreover, since we are dealing here with centrosymmetrical crystals, $\mathcal{G}'$ consists at least of the Identity 1 and the Inversion $\overline{1}$.

There are two ways to ensure that the reconstructed $p(\vec{r})$ will obey the symmetry requirements. The first one consists in enlarging artificially the measured set of reflections by generating all the unmeasured but equivalent reflections using Waser's formula (Waser 1955). If a single reflection (F, σ) is used to generate N equivalent reflections, the error bar attached to each one of these must now be set to $\sqrt{N}\sigma$, in accordance with (9). The major disadvantage of this procedure is to enlarge the size of the **A** matrix, which is proportional to the required computing time. This latter fact stems from the iterative nature of the Maximum Entropy algorithm, which involves above 20 iterates on the average and 6 matrix multiplications involving **A** per iterate.

---

† More magnetic Bragg peaks may exist, which do not coincide with nuclear Bragg peaks, e.g. in an antiferromagnet. In such a case, **only** the **induced paramagnetic part** of the magnetisation density will be reconstructed by our procedure.

A second and more efficient way makes use of group theory. It can be shown (Papoular 1990) that formula (2) from section 2.2 still holds if the exponential term is replaced by its average over all symmetry operations of $\mathcal{G}'$. It will be shown in section 3.2 that the Maximum Entropy algorithm reconstructs a suitably symmetrical projected magnetisation density $p(\vec{r})$ and only uses a set of symmetrically independent scattering vectors. Due to our restriction to centrosymmetrical crystals, the averaged exponential can be replaced by an averaged cosine.

More specifically, let $\hat{R}_n$ be a symmetry operation belonging to $\mathcal{G}'$. Let $\hat{\alpha}_n$ and $\vec{\beta}_n$ be the rotation and translation parts of $\hat{R}_n$.. The exponential in formulae (2) and (6b) is now to be replaced by:

$$\left\langle \exp\{i\vec{K} \cdot \vec{r}\} \right\rangle = \frac{1}{h'} \cdot \sum_{n=1}^{h'} \cos\{\vec{K} \cdot \hat{R}_n \vec{r}\} \quad \text{where:} \quad \hat{R}\vec{r} = \hat{\alpha}\vec{r} + \vec{\beta} \tag{10}$$

3. THE MAXIMUM ENTROPY SOLUTION

3.1 The Maximum Entropy recipe

Since the fundamentals of Maximum Entropy are by now well established and documented, the reader is referred to the literature for the underlying philosophy and mathematical justifications (Jaynes 1983, Frieden 1972a,b Tikochinsky *et al* 1984), Livesey and Skilling 1984), algorithms (Skilling and Bryan 1984, Skilling and Gull 1985), earlier spectacular successes (Frieden 1978, Gull and Skilling 1984) and most recent developments (Skilling and Gull 1989).

Maximum Entropy requires two master quantities to be introduced: $\chi^2\{p(\vec{r})\}$ and the Shannon-Jaynes entropy $S\{p(\vec{r})\}$ respectively defined by:

$$\chi^2\{p(\vec{r})\} = \sum_{k=1}^{N} \left[\frac{F_m(\vec{K}_k) - F_m^c(\vec{K}_k)}{\sigma(\vec{K}_k)} \right]^2 \tag{11a}$$

and

$$S\{p(\vec{r})\} = -\sum_{j=1}^{M} \{q_{\uparrow j} \cdot \ln q_{\uparrow j} + q_{\downarrow j} \cdot \ln q_{\downarrow j}\} \tag{11b}$$

where:

$$q_{\uparrow j} = \frac{p_\uparrow(\vec{r}_j)}{\sum\limits_{j=1}^{M} [p_\uparrow(\vec{r}_j) + p_\downarrow(\vec{r}_j)]} \quad \text{and} \quad q_{\downarrow j} = \frac{p_\downarrow(\vec{r}_j)}{\sum\limits_{j=1}^{M} [p_\uparrow(\vec{r}_j) + p_\downarrow(\vec{r}_j)]} \tag{11c}$$

In (11a), $\chi^2\{p(\vec{r})\}$ depends upon $p(\vec{r})$ via formulae (6a) and (6b).

It should be noted that, due to the occurence of the logarithm in the expression of the entropy, only positive distributions can be retrieved. Looking for a non-positive distribution by means of the difference of two positive distributions, which amounts to looking for a positive distribution of twice the initial size, has first been used successfully in N.M.R. data analysis.

The Maximum Entropy method results from a Bayesian analysis in which one aims at maximizing:

$$Prob(p(\vec{r})|Data) \propto Prob(Data|p(\vec{r})) \cdot Prob(p(\vec{r})) \qquad (12)$$

where $Prob(X|Y)$ is the conditional probability of getting X knowing Y, $Prob(X)$ is the prior probability of X and $p(\vec{r})$ is any 'feasible' spin density map obeying the symmetry requirements. In the above formula, $Prob(Data|p(\vec{r}))$ is the likelihood, and is proportional to $\exp\{-\frac{\chi^2}{2}\}$. $Prob(p(\vec{r}))$ is essentially equal to $\exp\left[\alpha S\{p(\vec{r})\}\right]$ where $S\{p(\vec{r})\}$ is the entropy of the spin density map $p(\vec{r})$ and α is a Lagrange multiplier, the value of which is determined by the condition $\chi^2 = N$, the number of symmetrically independent data points. Maximizing $Prob(p(\vec{r})|Data)$ amounts to maximizing its logarithm, namely the Lagrangian $\mathcal{L} = \alpha S - \frac{\chi^2}{2}$.

The Maximum Entropy implementation used in the present work makes use of the Historical MaxEnt algorithm ('$\chi^2 = N$', no error bars on the entropic map) and the MEMSYS1 Kernel code from Cambridge. We recall here that the algorithm is iterative, yields a unique solution and converges when $\vec{\nabla}\chi^2$ is parallel to $\vec{\nabla}S$.

3.2 The spatial symmetry of the maximum entropy reconstruction ‡

For the sake of simplicity, and without loss of generality, let us consider any positive definite spin density map $\{p(\vec{r}_j)\}$ obeying the spatial symmetry requirements. Consider any pixel j, centred at $\vec{r}_j$. By operating each of the symmetry operations belonging to the projected factor space group $\mathcal{G}'$, a set C_j of g_j' distinct but symmetrically related pixels is generated, where g_j' is a submultiple of the order h' of $\mathcal{G}'$. Let us label each set by one of its pixels, say j_0. In order to ensure that for each j pertaining to C_{j_0}, the constraint $I_j = p(\vec{r}_j) = p(\vec{r}_{j_0}) = I_{j_0}$ is obeyed, let us introduce the Lagrange multipliers $\mu_{j_0}^j$ and the related Lagrangian $\mathcal{L}'_{j_0}$ so that:

$$\mathcal{L}'_{j_0} = \sum_{\substack{j \in C_{j_0} \\ j \neq j_0}} \mu_{j_0}^j \{I_j - I_{j_0}\} \qquad (13)$$

The remaining of the proof goes as follows. An overall Lagrangian $\mathcal{L}'$ has to be added to $\mathcal{L}$, which incorporates as many Lagrangians $\mathcal{L}'_{j_0}$ as there are symmetrically independent pixels j_0. The full Lagrangian for our problem now reads:

---

‡ This section may be skipped on first reading.

$$\mathcal{L}_{total} = \mathcal{L} + \mathcal{L}'$$
$$= \mathcal{L} + \sum_{j_0} \mathcal{L}'_{j_0}$$
$$= \alpha S - \frac{\chi^2}{2} + \sum_{j_0} \sum_{\substack{j \in C_{j_0} \\ j \neq j_0}} \mu^j_{j_0} \{I_j - I_{j_0}\} \qquad (14)$$

where I_j stands for $p(\vec{r}_j)$.

The Maximum Entropy condition reads:

$$\frac{\partial \mathcal{L}_{total}}{\partial I_j} \equiv 0 \quad , \quad \forall I_j \qquad (15a)$$

Taking the average over all j's belonging to the same C_{j_0} yields:

$$\frac{1}{g_{j_0}'} \cdot \sum_{j \in C_{j_0}} \frac{\partial \mathcal{L}_{total}}{\partial I_j} \equiv 0 \quad , \quad \forall j \qquad (15b)$$

It can be checked that the Lagrange multipliers $\mu^j_{j_0}$'s are elimated by this procedure. The derivative of χ^2 with respect to I_j involves the matrix element $A(k, j)$ † which is thus automatically averaged over. The last step is to note that the average over all j's belonging to C_{j_0} is the same as the average over $\hat{R}(j_0)$ for all $\hat{R}$'s belonging to $\mathcal{G}'$. **Q.E.D.**

4. APPLICATIONS AND DISCUSSION

In order to check that our procedure to retrieve $p(\vec{r})$ can be trusted despite noisy and limited available data, a simulation based upon a real experiment, involving a positive and negative spin density, was first run. Two real data cases involving molecular magnetism and high-T_c superconductivity are then presented. The first example shows direct, model-free, evidence of a chemical bonding involving π-like orbitals in a molecular compound: the tanol-suberate. The second example involves $HoBa_2Cu_3O_7$ in its superconducting phase. In this latter case, Maximum Entropy wipes out all truncation effects due to the very limited data set.

4.1 Simulation (*cf. fig.1*).

Two point-like magnetic moments of magnitude $+1.25\mu_B$ as well as two much smaller magnetic moments of magnitude $-0.25\mu_B$ were located in the projected cell in a centrosymmetrical way. Three data sets were simulated, involving respectively 23, 72 and 262 reflections. Error bars were estimated from a real experiment, no noise was added to the Fourier components themselves and the Inverse Fourier transform was computed

---

† refer to section 2.3, eq. (6a) & (6b).

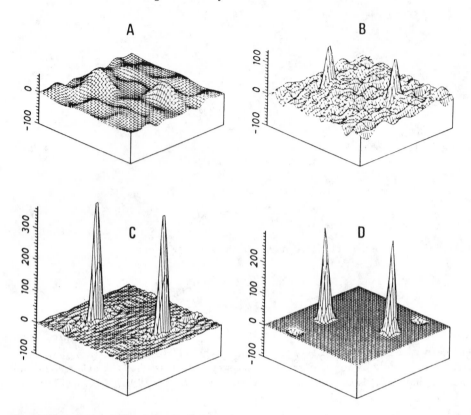

Fig. 1. Simulation involving two positive $(+1.25\mu_B)$ and two negative $(-0.25\mu_B)$ magnetic moments. The unit cell is assumed to be orthorhombic, and the only symmetry operations are the Identity and the Inversion. The projection is taken along the $\vec{a}$ crystallographic axis. The reconstructed magnetisation density is expressed in $\mu_B/\overset{\circ}{A}{}^2$. The size of the image is 50*50 pixels.

Orthographic projections:

A,B,C : straightforward Inverse Fourier transforms for an increasing number of reflections (23,72, and 262 respectively). Using more Fourier components does not eliminate truncation effects.

D: Maximum Entropy reconstruction using the same 23 reflections as in A. Truncation effects have vanished and the two negative features are now clearly apparent.

for all three cases. The results are shown on Fig.1 (A,B,C). Since no noise ‡ was taken into account, Fig.1 shows clearly that using more and more Fourier components does not eliminate truncation effects. By contrast, we have run Maximum Entropy on the smallest data set (23 Fourier components), yielding Fig.1(D). The 4 features of the original image are clearly restored. Moreover, the integrated values of each peak and each hole yielded $+1.19\mu_B$ and $-0.18\mu_B$ respectively. That the $\delta(\vec{r})$-peaks are broadened results from the limited set of data and from the error bars: Maximum Entropy produces the smoothest Image compatible with those. The overall integrated projected density yields $F(\vec{0}) = F(000)$, which is the averaged magnetization density over the unit cell. In the above example, one finds $2.02\mu_B$ instead of $2.00\mu_B$.

4.2 A first real data case: chemical bonding in tanol suberate (*cf. fig.2,3*).

This first real data case pertains to molecular magnetism. The motivation beyond this study is to understand the chemistry of organic free radicals, in this case a nitroxide biradical, and in particular to check the predictions from the theory that: i) the free electron should be localized on the NO group and ii) one expects a π-antibonding molecular orbital in the nitroxide group NO (Gillon 1983, Gillon and Schweizer 1989). Two different P.N.D. data sets were obtained, corresponding to the projection along the $\vec{a}$ axis and to the projection along the $(\overline{4}3\overline{3})$ direction.
69 independent Fourier components were measured in the first instance, but only 17 could be obtained in the second one. The space group for the tanol suberate is $P2_1/c$ and the number of symmetry operations compatible with the two projections mentioned above are respectively 4 and 2. The unit cell consists of two molecules of tanol suberate, each of which possesses two NO groups.

Let us discuss the projection along the $\vec{a}$ axis first (*cf. fig.2*). Whereas the standard Fourier analysis (Fig.2: A,C) could not resolve the electronic density about the NO group, the Maximum Entropy procedure yielded clearly resolved features(Fig.2: B,D). Moreover, the averaged magnetisation estimated from MaxEnt yielded $3.7\mu_B$ per unit cell against $3.2 \pm 0.2\mu_B$ as obtained from an independent macroscopic magnetisation measurement. The projection (Fig.2: D) should be understood as follows: Two p_z-orbitals are centred respectively on the Nitrogen and the Oxygen atoms from the nitroxide group. The direction of the projection is such that both a lobe of the p_z orbital centred at O and a lobe of the p_z orbital centred at N project approximately at the same place. That the two outside lobes have about the same intensity is a direct confirmation that the two orbitals have about the same weight within the NO bond. A ratio of 1.2:1 is found using two distinct sophisticated model refinements (Gillon 1983).

Let us consider now the second projection, along the $(\overline{4}3\overline{3})$ direction (*cf. fig.3*). The importance of this projection stems from the fact that its direction is almost perpendicular to the plane π of the NO bond. Due to the internal arrangement inside the unit cell, only one NO group out of two lies in a given plane π. The entropic estimate in this second case yielded $2.47\mu_B$, which is much lower than the expected value of $3.2\mu_B$, expressing the fact that the 17 independent Fourier components attainable in

---

‡ besides round-off computation errors.

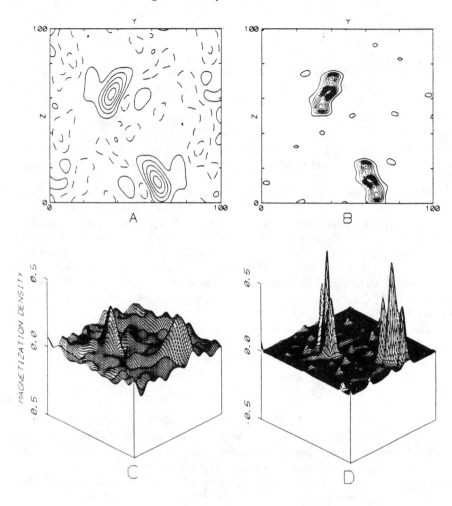

Fig. 2. A real data case: tanol suberate. The magnetisation density is projected along the $\vec{a}$ axis. The data consist of 69 independent Fourier components. The size of reconstructed image is 100*100 pixels.

Above, contour maps. A: Fourier, B: Maximum Entropy. The contour levels are the same in both A and B. The dotted lines refer to negative contours. Levels are $\pm.03, \pm.09, \pm.15, \pm.21 \ldots \mu_B/\mathring{A}^2$. The Maximum Entropy reconstruction clearly resolves the density about the Oxygen and Nitrogen atoms, whereas the Fourier reconstruction does not.

Below, 3-D orthographic projections: C: Fourier, D: Maximum Entropy. The Maximum entropy reconstruction shows that the weight of the p_z orbital for the Oxygen is close to that of the p_z orbital for the Nitrogen. The largest peak results from the overlap of the projections of both orbitals.

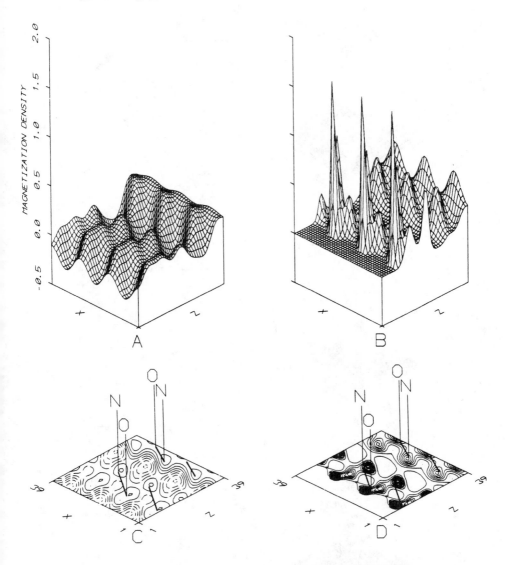

Fig. 3. Tanol suberate (cont'd): Projection of the magnetisation density along the ($\bar{4}33$) direction. The data consist of 17 independent Fourier components. The size of the reconstructed image is 39*39 pixels.

Above, 3-D orthographic projections. A: Fourier B: Maximum Entropy

Below, contour maps. C: Fourier D: Maximum Entropy

The contours are the same as in Fig.2. The Maximum Entropy reconstruction reveals the antibonding character of the NO molecular orbital involving p_z-type orbitals on both the Oxygen and the Nitrogen. The straight solid line connects two atoms from the same molecule of tanol suberate.

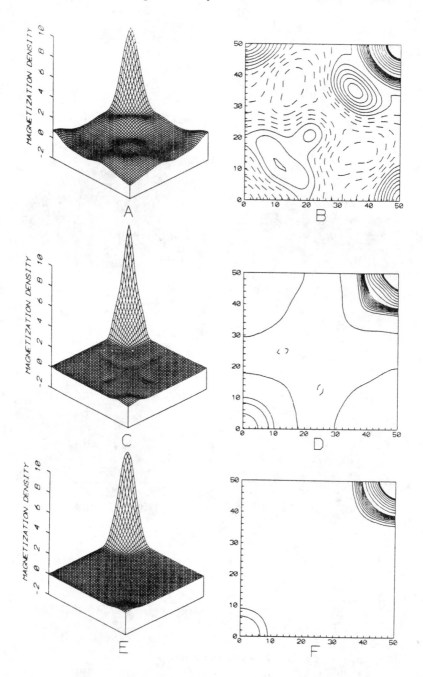

Fig. 4. Superconducting $HoBa_2Cu_3O_7$ at T = 1.7 Kelvin and H = 5 Tesla. The projection is taken along the $\vec{c}$ axis. The data consist of 21 Fourier components.
A,B: Fourier map - C,D: Positive/negative MaxEnt map - E,F: Positive MaxEnt map

that case were not enough, together with the symmetry requirements, to incorporate the whole information. The antibonding character is indeed straightforwardly revealed by Maximum Entropy, as evidenced in Fig.3 D. In this figure, an Oxygen and a Nitrogen belonging to the same molecule are connected by a straight solid line. Note the two distinct types of projected NO bonds.

4.3 A second real data case: Superconducting high-T_c $HoBa_2Cu_3O_7$ at T = 1.7 Kelvin and H = 5 Tesla (*cf. fig. 4*).

As a second example, we consider a case where the data are very scarce. Only 21 independent Fourier components are measured (Gillon *et al* 1990). The magnetisation density is projected along the $\vec{c}$ axis. The Fourier transform (A,B) produces huge truncation effects. In the 3-D orthographic projections (A,C,E), the magnetic Holmium atom projects onto the upper corner whereas the $Cu(1)$, $Cu(2)$ and $O(1)$ atoms project onto the lower corner. $O(2)$, $O(3)$,$O(4)$ atoms project onto the side corners. The truncation effects mentioned above preclude from drawing any conclusion. As a matter of fact, the side small peaks are entirely artefacts, as shown by the entropic reconstructions. Two distinct entropic reconstructions involving either a positive/negative or a positive definite magnetisation density are shown in (C,D) and (E,F). The positive/negative map shows some negative density close to high symmetry points in projected cell. Although most probably due to truncation or noise leakage since these positions do not correspond to any atom within the unit cell, assigning error bars to these effects would be desirable. The common features of the two entropic reconstructions are: i) no evidence for magnetic moments on $O(2)$,$O(3)$ and $O(4)$ atoms ii) a definite evidence for a small contribution on either the $Cu(1)$,$Cu(2)$ or $O(1)$ atom. It is the purpose of our future work to try estimate error bars on such features using Quantified Maximum Entropy (Skilling et Gull, 1989). If real, these are crucial to the understanding of the high-T_c systems.

5. CONCLUSIONS

The Maximum Entropy reconstruction technique has been successfully applied to Polarised Neutron Diffraction on centrosymmetrical crystals, which is essentially a Real Fourier method. A key requirement is that the desired image must retain the spatial symmetry of the crystal. We have shown that this can be achieved by suitably symmetrizing the integrand of the Fourier transform and considering only symmetrically independent magnetic structure factors. A substantial gain in time results, because of the iterative nature of the Maximum Entropy algorithm.

Both simulations and real data cases were studied, showing that Maximum Entropy is very promising for the case when data is scarce and noisy, as in high-Tc superconductors. This should prove to be even more so, thanks to a new MaxEnt procedure (Quantified MaxEnt) which allows for error bar estimations on magnetisation density peaks (Skilling and Gull 1989). This huge improvement, soon to be applied to our case, will help us to infer more from our Polarised Neutron Diffraction experiments.

The two conventional image-reconstruction techniques in current use have serious drawbacks. The first one, based on a smoothed inverse Fourier transform, is plagued with heavy truncation effects. The latter, Least-Squares, introduces a strong bias via an a priori model. Running Maximum Entropy first, which alone provides a direct look at the spatial magnetization density, should help us to infer a best suited (and least biased) model.

Last but not least, provided that enough data can be gathered, Maximum Entropy yields a fairly good approximation of the averaged magnetization per unit cell without having recourse to a direct measurement or a model.

ACKOWLEDGEMENTS

The authors wish to thank M. Lambert and G. Jannink for support and encouragement. They are very much indebted to J. Schweizer (CEN/G), F. Tasset (ILL) and J. Skilling (DAMTP) for highly stimulating discussions and suggestions. One of us (R.J.P.) wishes to thank J. Skilling and S. Gull (Cambridge, UK) for the use of their MEMSYS1 package.

REFERENCES

Brown P J, Forsyth J B and Mason R 1980 *Phil. Trans. R. Soc. London.* **B 290** 481
Burg J P 1967 *Ann. Meet. Int. Soc. Explor. Geophys.* Reprinted in *Modern Spectral Analysis* 1978 ed D G Childers (New-York:IEEE Press) pp 34-41
Frieden B R 1972 *J. Opt. Soc. Am.* **62** 511
Frieden B R and Burke J J 1972 *J. Opt. Soc. Am.* **62** 1202
Frieden B R and Swindell W 1976 *Science* **191** 1237
Frieden B R and Wells D C 1978 *J. Opt. Soc. Am.* **68** 93
Gillon B 1983 *Thèse d'Etat, Orsay, France*
Gillon B and Schweizer J 1989 in *Molecules in Physics, Chemistry and Biology.*
    Vol III ed J Maruani (Kluwer) pp 111-147
Gillon B, Spasojevic-de Biré A, Petitgrand D, P Schweiss and Collin G 1990 *Proc.*
    XVth Congress of the International Union of Crystallography, Bordeaux, France
Gull S F and Daniell G J 1978 *Nature* **272** 686
Gull S F and Skilling J 1984 *IEE Proceedings* **131**F 646
Jaynes E T 1983 *Papers on Probability, Statistics and Statistical Physics*
    ed R D Rosenkrantz (Kluwer)
Johnson M W ed 1986 *Neutron Scattering Data Analysis 1986*
    IOP Conference Series (Bristol) **81**
Laue E D, Skilling J, Staunton J 1985 *J. Magn. Reson.* **63** 418
Laue E D, Mayger M R, Skilling J and Staunton J 1986 *J. Magn. Reson.* **68** 14
Livesey A K and Skilling J 1985 *Acta Cryst.* **A 41** 113
Narayan R and Nityananda R 1986 *Ann. Rev. Astron. Astrophys.* **24** 127
Papoular R J 1990 *Preprint LLB/18* (Submitted to *Acta Cryst. A*)
Shull C G and Mook H A 1966 *Phys. Rev. Lett.* **16** 186
Skilling J and Bryan R K 1984 *Mon. Not. R. Astr. Soc.* **211** 111
Skilling J and Gull S F 1985 in *Maximum Entropy and Bayesian Methods in Inverse Problems.* ed C Ray Smith and W T Grandy Jr. (D. Reidel) pp 83-132
Skilling J and Gull S F 1989 in *Maximum Entropy and Bayesian Methods* ed J Skilling (Kluwer) pp 45-52 and 53-71
Tikochinsky Y, Tishby N Z and Levine R D 1984 *Phys. Rev. Lett.* **52** 1357
Waser J 1955 *Acta Cryst.* **8** 595

Inst. Phys. Conf. Ser. No 107: Chapter 2
Paper presented at Neutron Scatt. Data Anal. Conference, Rutherford Appleton, 1990

117

Real-time neutron powder diffraction study of $MoO_3.2H_2O$ topotactic dehydration

M. Anne[a], N. Boudjada[a], J. Rodriguez[b] and M. Figlarz[c]

(a) Laboratoire de Cristallographie, C.N.R.S., associé à l'Université J. Fourier, 166X, 38042 Grenoble Cedex, France.
(b) Institut Laue Langevin, 156X, 38042 Grenoble Cedex, France.
(c) Laboratoire de Réactivité et de Chimie des Solides, Université d'Amiens, 80039 Amiens Cedex, France.

ABSTRACT: Experience gained during recent years in real-time neutron diffraction with powder diffractometers has shown that thermodiffractometry is a fast and powerful method to obtain both structural and kinetics information on solid state transformations in one single experiment, and thus supplements more conventional techniques. This is especially true in the case of reactions involving an evolution of protonated species: changes in the incoherent background can then be used to monitor the sample composition.

Among the various molybdic acids quoted in the literature, two compounds with the formulae $MoO_3.2H_2O$ and $MoO_3.H_2O$ are definite phases and have been extensively studied by X-ray single crystal diffraction and NMR. In the present work, a neutron thermodiffractometric study was carried out, in order to investigate the topotactic dehydration of $MoO_3.2H_2O$, leading to $MoO_3.H_2O$ and MoO_3.

1.   INTRODUCTION

Studies on the mechanism of solid state reactions are largely based on single crystal studies by X-ray or electron diffraction. Unfortunately such experiments do not provide any information about the kinetics of the process and, when kinetics data become available, e.g. through calorimetric or TGA (thermogravimetric analysis) investigation, they cannot usually be easily correlated with the diffraction data (for instance because of the large differences of sample environment which may strongly influence the transformation itself).

The current development of real time neutron powder diffraction (RTNPD) is the result of the conjunction of new and efficient tools: fast computers with large memories, powder diffractometers which can give a full pattern within one minute, together with the fact that in one single experiment a powder diffraction pattern gives a lot of information concerning both the structure (atomic positions) and the morphology (size and shape of the crystallites) of a crystalline solid.

In the following, first the software package STRAP (a System for Time-Resolved data Analysis, Powder diffraction patterns) is described, and then its application to a specific study concerning the topotactic dehydration of molybdenum trioxide dihydrate is presented.

2.   NEUTRON POWDER DIFFRACTION DATA ANALYSIS

Many compounds cannot be prepared as single crystals suitable for diffraction measurements. This is especially true for neutron diffraction, which requires samples larger than those needed for X-ray analysis. Even in the case when it is possible to grow large single crystals, these may still suffer from such effects as magnetic domain structures, phase transitions involving twin formation or dimensional changes which may pulverize the starting material. In these cases, the use of powders is the only possible choice. The major problem in using powder diffraction is the overlap of diffraction lines: all reflections which scatter at a given Bragg angle are bunched together. The first step of any powder data analysis will then be to separate the contributions from the various hkl planes in order to extract crystallographic information. The final stage of the analysis will be the determination, if possible, of the crystal structure, which can be achieved using for instance the well known full profile refinement procedure firstly introduced by Rietveld (1967, 1969).

2.1   Real time neutron powder diffraction

The use of PSD (Position Sensitive Detectors)-equipped powder diffractometers and a high neutron flux make it possible to follow in real time the evolution of physical and chemical properties of compounds by neutron diffraction (Pannetier 1986). The number of applications is very large: kinetics of chemical reactions, phase transitions (Rodriguez *et al* 1987b, 1988) related to the variation of thermodynamic parameters such as temperature (thermodiffractometry), pressure, magnetic field, in situ electrochemical intercalation (Poinsignon *et al* 1989), etc...

In addition, in the case of hydration or dehydration of solids, the high incoherent background due to hydrogen makes it possible to investigate the proton content and structural properties of a sample simultaneously.

An obvious limiting condition for a time resolved study is the necessity of a much shorter acquisition time for a full diffraction pattern (with enough statistical accuracy) than the characteristic time of the observed phenomenon. On the other hand, the accumulation of a few hundred complete diffraction patterns per day requires well adapted programs to analyse these data.

2.2 The set of programs STRAP

The software package STRAP developped at I.L.L. (Rodriguez *et al* 1987a) is a set of programs which is a system of data analysis and plotting for Neutron Powder Diffraction - NPD - data produced by the ILL powder diffractometers D1A, D1B, D2B, and D20 (constant wavelength, angle dispersive diffractometers). This system is available on the VAX 8650, 8700 central computer and can be run on any visitor zone of the above instruments.

Three levels of data analysis are available:

• numerical analysis: this is the option to use when no a priori information on the pattern is available (unknown cell parameters for instance) or when the pattern cannot be described in the conventional crystallographic approach (e.g. quasicrystals). In such a case, the analysis will be restricted to fitting a given line shape(s) above a polynomial background. For a unknown crystalline compound, the next step would normally be the determination of the cell parameters by using an automatic indexing method.

• cell constrained analysis (Pawley-type program): this approach is used to extract refined cell parameters and intensities from NPD patterns. The only information which has to be fed into the program is the symmetry (Laue group) and approximate cell parameters. This program would normally be the first step of structure determination from NPD data by direct methods.

• profile fitting structure refinement (Rietveld-type refinement): this is the most common procedure when the structure of the compound is at least approximately known. This method will fit the diffraction pattern using the instrumental characteristics (resolution curve) and structural parameters of the sample (cell parameters, atomic and thermal parameters) as variables.

All three levels of data analysis have already been discussed at length in the literature and the system STRAP, which is largely based on published and well documented routines does not bring anything new to the mathematical aspects of NPD data analysis. The emphasis of this system is rather on the analysis of sequential sets of data as generated by real time diffraction experiments on PSD diffractometers. A characteristic common to all time-resolved experiments (e.g. kinetic studies, thermodiffractometries...) is indeed the huge amount of patterns (typically 10 to 50 patterns per hour on D1B) recorded in a single experiment.

Analysis of all these patterns by conventional methods (i.e. one pattern at a time) is clearly impractical and often impossible. However, the high rate of data acquisition usually means that a given pattern differs only very little from the previous or following one; in other words, the (refined) parameters which characterize a given diffraction pattern provide a good approximation to start refining the parameters of the next pattern. STRAP has been designed to make such a sequential analysis of data as convenient and automatic as possible.

Of course refinement of single neutron (or X-ray) powder diffraction patterns is also possible with STRAP.

Various plotting programs are available to prepare either rapid or high quality plots of the raw data. Two kinds of representation can be used:

• 2D-plots for single patterns: produces drawings of $I=f(p)$ where $p=2\theta$, d or $q(=4\pi\sin\theta/\lambda)$. This is obviously the most convenient and quickest way to examine a diffraction pattern and to assess the quality of both sample and experimental conditions.

• 3D-plots and/or contour plots for sequences of NPD patterns: the result is a surface representation of the function $I=f(2\theta,p)$ where p=pattern running number,temperature,time, cradle angles,voltmeter values, etc... These plots are particularly valuable to illustrate real-time diffraction experiments but can also be used to stack a limited set of patterns on top of each other for comparison.

3.   TOPOTACTIC DEHYDRATION OF $MoO_3. 2H_2O$

A example of real-time application of neutron powder diffraction is the study of the topotactic dehydration of $MoO_3. 2H_2O$ by neutron thermodiffractometry.

Molybdenum trioxide hydrates or molybdic acids have often been mentioned in the literature and belong to the interesting family of fast proton conducting solids. Günter (1972) has shown that thermal dehydration of molybdenum trioxide dihydrate proceeds in two steps. The transformation is topotactic and the resulting compounds are definite phases which have been studied by X-ray single crystal diffraction (Lindquist 1950, 1956, Krebs 1972, Cesbron 1985) and NMR (Jarman 1981). The dehydration reaction is:

$$MoO_3. 2H_2O \rightarrow MoO_3. H_2O \rightarrow MoO_3.$$

The structure of the dihydrate $MoO_3. 2H_2O$ is built from slightly distorted layers of corner sharing $MoO_5(H_2O)$ octahedra; these layers are stacked above each other and connected by interlayer water molecules (Figure 1). Although the relation between the three structures has been clearly determined by single crystal X-ray photographs (Günter 1972), the recent developments of real time neutron powder diffraction have made it possible to obtain information on the mechanisms driving the topotactic dehydration of $MoO_3. 2H_2O$.

Preliminary results of a recent thermodiffractometric study are given below. A more detailed and quantitative interpretation will be presented in a forthcoming article (Boudjada 1990).

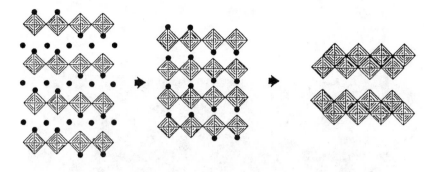

<table>
<tr><td>(2)
$MoO_3 . 2H_2O$</td><td>(1)
$MoO_3 . H_2O$</td><td>(0)
MoO_3</td></tr>
</table>

(2) $MoO_3 . 2H_2O$	(1) $MoO_3 . H_2O$	(0) MoO_3
a = 10.476	a = 7.55	a = 3.96 Å
b = 13.822	b = 10.69	b = 13.86 Å
c = 10.606	c = 7.28	c = 3.70 Å
β = 91.62 °	β = 91 °	
$P2_1/n$	$P2_1/c$	Pmnb

$$[101]_2 \ // \ [100]_1 \ // \ [001]_0$$
$$[010]_2 \ // \ [010]_1 \ // \ [010]_0$$
$$[-101]_2 \ // \ [001]_1 \ // \ [100]_0$$

Figure 1.   Comparison of the schematic structures involved in the topotactic dehydration of $MoO_3 . 2H_2O$. The b-axis is perpendicular to the MoO_6 sheets. Black circles are H_2O.

3.1 Experimental

The thermodiffractometric experiment was carried out on the high flux PSD diffractometer D1B. The wavelength was λ = 2.52 Å; the sample (powdered $MoO_3 . 2H_2O$ in a vanadium cylinder: h = 50 mm, Φ = 10 mm) was heated in a furnace under vacuum from RT to 400 °C at a rate of 20 °C / h. Diffraction patterns were collected every 3 mn and the temperature change was 1 °C per pattern. A total of 359 diffraction patterns have been analysed sequentially, using the Rietveld method of full profile refinement.

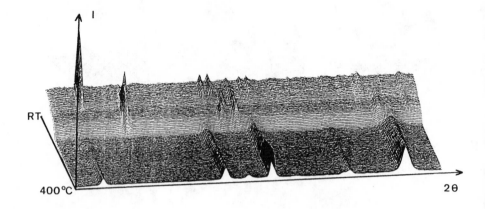

Figure 2.   Temperature evolution of the powder pattern of $MoO_3 \cdot 2H_2O$, leading to $MoO_3 \cdot H_2O$ and MoO_3. The background is approximately proportional to the water content.

3.2   Results

A 3D plot of diffraction data from the whole temperature range is displayed on Figure 2. The intensity of the incoherent background is essentially proportional to the amount of hydrogen in the sample; its decrease in two steps, which is observed on heating, corresponds to the variation in the number of water molecules from one phase to the other, and materializes the range of existence of the three phases. Furthermore, the relatively smooth evolution of the Bragg's peaks, together with the background intensity, is explained by the co-existence of two phases in the temperature range where the transformation occurs. The diffracted lines are very broad when the compound becomes anhydrous (MoO_3) and then narrow when the temperature increases; this is an indication of amorphousness and re-crystallization at the last step of dehydration.

The curves plotted on Figure 3 represent the evolution with the temperature of cell parameters and volume, scale factors and selected normalized intensities for the three phases.In the case of the dehydrated phase MoO_3, the average crystallite size, normal to various (hkl) planes was calculated from the profile parameters of five selected peaks and represented as a function of the temperature on Figure 4.

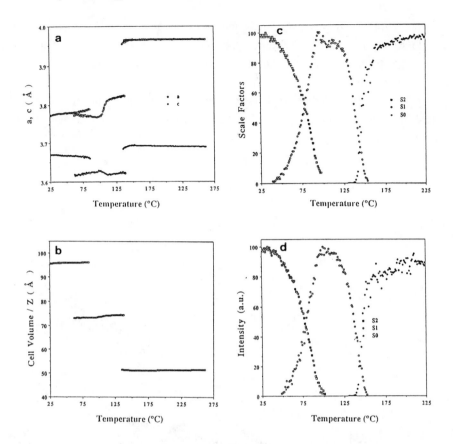

Figure 3.   Temperature evolution of -a) the reduced cell parameters of MoO_3. nH_2O (n=2,1,0) -b) the volumes reduced to one formula unit -c) the normalized scale factors and -d) the integrated and normalized intensities of selected reflections.

3.3   Discussion

The cell parameters are reduced, in order to be compared to those of MoO_3 (Figure 3a). Leaving aside the b-axis, which is perpendicular to the MoO_6 sheets, the a- and c- parameters vary in opposite ways in the dihydrate, whereas they remain nearly constant in MoO_3 when the temperature is increasing. In the case of the monohydrate, the variation of the cell parameters shows an anomalous behaviour around 95 °C, which is characteristic of a structural phase transition. This anomaly is also observed in the scale factor evolution of the monohydrate phase, on Figure 3c.

The reduced volume (volume per formula unit) decreases about 25 Å^3 for each of the two steps of the dehydration process (Figure 3b); both steps correspond to the loss of one water molecule per formula unit. The reactions are comparable for the two stages and have occurred throughout the full volume of the cell.

The evolution of the normalized scale factors (Figure 3c), as well as the evolution of the selected intensities from the three phases (Figure 3d), confirm the co-existence of two phases at each transformation.

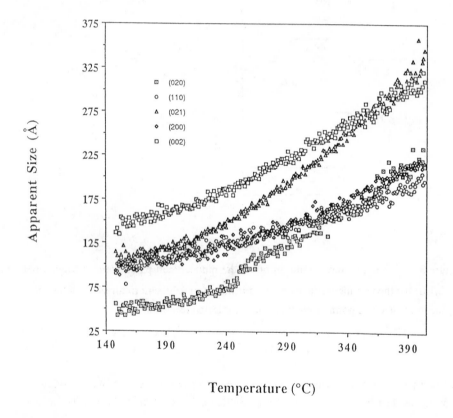

Figure 4.   Temperature evolution of the "apparent" crystallite size of MoO$_3$, along five selected directions.

The average crystallite sizes have been calculated in several directions for MoO_3, from the profile parameters of selected reflections. The pseudo-Voigt function (Langford *et al* 1986) was used to describe the peak shape and an "apparent" crystallite size was deduced from Scherrer's formula:

$$<D>_{app.} = \lambda/\beta \sin\theta$$

where β is the integral breadth of the unit area normalized pseudo-Voigt function. It is clear, on Figure 4, that the preferred direction of crystal growth is along [210]; at the early stages of the formation of MoO_3, the crystallites have a marked anisotropy.

The dehydration reactions seem to begin in a part of the sample volume and stretch through the whole volume to crystallize the new phase.

4. CONCLUSION

The study of the topotactic dehydration of molybdenum trioxide dihydrate was an interesting example to show all the information which can be obtained from an RTNPD experiment, using the system STRAP to analyse the data. The topotactic character of the dehydration has been verified by sequentially refining the cell parameters, the incoherent background has been related to the chemical composition of the sample, and finally, details have been given, from the analysis of the sample broadening, about the re-crystallization of the dehydrated phase MoO_3.

REFERENCES

Boudjada N, Rodriguez-Carvajal J, Anne M and Figlarz M 1990 *to be published*

Cesbron F and Ginderow D 1985 *Bull. Mineral.* **108** 813

Günter J R 1972 *J. Solid State Chem.* **5** 354

Jarman R H, Dickens P G and Slade R C T 1981*J. Solid State Chem.* **39** 387

Krebs B 1972 *Acta Cryst.* **B28** 2222

Langford J L, Louër D, Sonneveld E J and Visser J W 1986 *Powder Diffraction* **1**(3) 211

Lindquist I 1950 *Acta Chem. Scand.* **4** 650

Lindquist I 1956 *Acta Chem. Scand.* **10** 1362

Pannetier J 1986 *Chemica Scripta* **26A** 131

Poinsignon C, Forestier M, Anne M, Fruchart D, Miraglia S, Rouault A and Pannetier J 1989 *Zeit. Physik. Chem. N. F.* **164** 1515

Rietveld H M 1967 *Acta Cryst.* **22** 151

Rietveld H M 1969 *J. Applied Cryst.* **2** 65

Rodriguez J, Anne M and Pannetier J 1987a STRAP *ILL Internal Report* 87RO14T

Rodriguez J, Gonzalez-Calbet J M, Grenier J C, Pannetier J and Anne M
  1987b *Solid State Comm.* **62** 231

Rodriguez J, Bassas J, Obradors X, Vallet M, Calbet J, Anne M and Pannetier J
  1988 *Physica C* **153-155** 1671

MXD: A least-squares program for non-standard crystallographic fitting

by

P WOLFERS

Laboratoire de Cristallographie, associé à 'Université J Fourier
166X, 38042 Grenoble, France

MXD is a new crystallographic least-squares program designed for a wide range of structure determinations, including complex magnetic and/or modulated structures with any number of wave vectors. The data to be analysed may be from X-rays, neutrons or polarised neutrons, with powder, single crystal or twinned samples.

INTRODUCTION

The use of neutron diffraction to study the magnetic properties of 3d metal compounds leads to the determination of complex magnetic structures. Several studies [1-6] have shown that it is necessary to have a generalised refinement program to obtain the necessary fit to the resultant magnetic structure. With powder data, since the list of measured reflections may be short, it is especially important that all possible constraints be introduced in order to reduce the number of independent parameters.

These can be :

- The symmetry constraints as obtained from the macroscopic theory of Bertaut [7]
- Constraints to ensure the same magnetic moments magnitude on equivalent sites
- Constraints to ensure the resulting magnetisation is in agreement with magnetisation measurements

In 1982, we received the NBS Pascal compiler for a PDP11 [8] (Digital Equipment Corp) systems. This compiler, written in Pascal, uses an original algorithm with memory object trees that generate the processor code (Figure 1).

Using this technique, we have written a program module to manage any expression in least-squares processing where the partial derivates of any defined function can be computed directly. This "formula manager" is the base of the MXD program. It gives a new approach in the design of the control files, where the constraints are replaced by the formula dependence concept.

Memory Tree Example

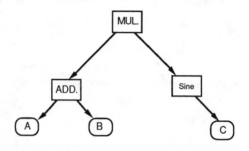

Formula : (A + B) * Sin(c)

Figure 1

If you imagine an atom lying along the three–fold axis of a cubit unit cell, with all current crystallographic programs, you write something like this :

Traditional program :

ATOM Fe3   0.273, 0.273, 0.273

The program, by the use of given symmetry elements deduces that the fitted parameter x (set to 0.273) is linked to y and z parameters by the equality relation $x = y = z$. This is the traditional method of introducing the constraint.

With MXD, you define the variable parameters to be refined (so called the "least–squares variables") by a statement, and after, you define your atom :

MXD :

Variable xfe3 = 0.273;

ATOM 'Fe3' xfe3, xfe3, xfe3;

If you want to define a magnetic moment in spherical coordinates on this atom, you can write :

A magnetic moment in spherical coordinates :

Variable	M = 4.8	(in Bohr magneton units)
	Th = 42.5	(theta angle in degrees)
	Ph = 24	(phi angle)

Moment	'Fe3'	M*Sin(Th)*Cos(Ph)	(Mx)
		M*Sin(Th)*Sin(Ph)	(My)
		M*Cos(Th)	(Mz)

If you want to define some form factor, you can use interpolated value tables of formulae. For example, the Pepinsky formulae can be expressed as :

Form factor as formula :

$$f = a_1.e^{-(h/b_1)^2} + a_2.e^{-(h/b_2)^2} + ...$$

Example for two exponential functions :

Param	A1 = 1.237, B1 = 0.5839,
	A2 = -0.237, B2 = 0.4442

Param	Fe3frm = A1*EXP(-(($sithsl/B1)**2))
	+ A2*EXP(-(($sithsl/B2)**2))

where $sithsl is the sin(theta)/lambda of the current reflection.

As shown by this example, MXD enables the user to describe his particular problem, and to perform an efficient least–squares refinement without writing a program.

We can summarise the main features of MXD :

The MXD capabilities :

A. Data elements can be one of the following :

1. A reflection line intensity for powder, twin or magnetic domains (type Int)
2. An observed F^2 (squared structure factor)
3. An observed F structure factor magnitude

B. Each reflection line of any type (Int, F^2 or F) can eventually be associated with a wave vector and/or a neutron polarisation direction.

1. MXD can simultaneously use any number of data of any of the above types.
2. Any number of wave vectors can be specified for modulated magnetic structure an/or for charge density wave.

C. MXD can take into account twin or magnetic domains in a crystal sample

D. Interface programs already exist allowing the exchange of data from and to other crystallographic systems.

A TYPICAL MXD APPLICATION

Let us illustrate with the study of the super–magnet–like compound $Ho_2Fe_{14}B$ [9,10], a typical MXD application.

This compound exhibits a magnetisation rotation of 20° from the normal $\bar{c}$ direction, the four fold axis, in the [110] plane of the tetragonal unit cell. Table 1 summarises some typical $Ho_2Fe_{14}B$ data. The spin reorientation takes place below 58K.

TABLE I

$Ho_2Fe_{14}B$

Space Group : P 4_2/mnm

Unit cell : a = 8.792 Å, c = 12.177 Å, (z = 4).

Crystallographic sites :

Ho_1	4f	(x, x, 0)	Ho_2	4g	(x, x, 0)
Fe_1	4c	(0, 1/2, 0)	Fe_4	8j	(x, x, z)
Fe_2	16k	(x, y, z)	Fe_5	8j	(x, x, z)
Fe_3	16k	(x, y, z)	Fe_6	4c	(0, 0, z)
B	4f	(x, x, 0).			

Spin rotation of 20° below 58 K.

Magnetic moments :

Ho	-10 μ_B,
Fe	From 1.9 to 3.4 μ_B.

Although the room temperature colinear magnetic structure is well known [10], the low temperature structure of this compound was unknown. So we have undertaken the determination of this magnetic structure with a spherical single crystal of 1.6 mm in diameter. The experiment was performed on the D9 four circle neutron diffractometer at the ILL.

Data Corrections

1. When the spin rotation appears, the maximum possible symmetry is the space group C2/m, so we have to obtain eight different magnetic domains (Figure 2) because we have not applied a magnetic field on the sample. These different domains can be deduced from each other by application of the factor group 4/m. In fact, opposite domains give the same intensities, even if the symmetry centre is lost, because the Fermi length of B is very close to a real value. We have therefore limited the number of domains to four, and each measured intensity was computed as the sum of the four domain contributions (Figure 2).

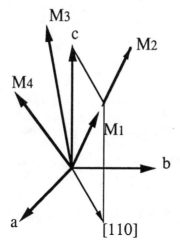

The four easy magnetization directions

Figure 2

To do so, we use the powder mode of MXD where each intensity is computed as the sum of a set of several specified reflections.

$$I(hkl) = p_1 {}^*I(h,k,l) + p_2 {}^*I(-k,h,l) + p_3 {}^*I(-k,h,l) + p_4 {}^*I(-k,h,l)$$

2. The second important correction was due to the secondary extinction that weakened some reflections by a factor of 0.75. We have used the Becker and Coppens correction [11] on a crystal type I with a mosaic Lorentzian distribution.

3. The last correction is small. We have noticed that the measured reflection pair (a reflection and its antisymmetric one) was not perfectly equal at low temperature. Since the sample had been slightly deviated from the confusion sphere of the diffractometer, the inhomogeneous beam, extracted by a focusing monochromator, played a role in the final diffracted intensity distribution.

We have corrected this effect by using an adjustable reflection dependent scale factor, defined by the next formula :

$$Dcorr = 1 + <Gf|Dc> + <Dc \mid Tf \mid Dc>$$

Dc is the deviation vector (diff coordinates)
Gf is the neutron flux gradient
Tf is a secondary order tensor

Dc is expressed in terms of the diffractometer coordinates, as functions of the goniometer Eulerian angles.

4. As the anisotropy of the iron is weak, we have forced all iron magnetic moments to be parallel to the inverse of the resulting sum of the rare earth moments.

Results Summary

The resulting structure is being published [12]. The non–colinear magnetic structure determined by this method reveals a negative exchange interaction between the rare earth atoms, and a large distortion of the crystallographic structure. The forbidden diffraction lines of the tetragonal space group are observed with substantial nuclear contributions. These results are in good agreement with the ^{57}Fe Mössbauer of the Hirosawa group [13]. The true symmetry is probably Cm up to the Curie temperature.

PARTICULAR MXD FEATURES USED IN THIS APPLICATION

The study of $Ho_2Fe_{14}B$ has required :

- Twin and/or magnetic domain management
- Linex–like secondary extinction (Becker and Coppens)
- Special data correction
- Special constraints

CONCLUSION

MXD is a general least–squares program that has been used with success by an increasing number of crystallographers, without known bugs. It can be run on any VAX/VMS (DEC) system and in a few months on any IBM–PC computer with floating point processor. The main fields where MXD use can be decisive are summarised in Table 2.

TABLE II

Main MXD application field

Any combination of the next items :

Data	X–ray and/or neutrons and/or polarized neutrons
Sample(s)	Powder and/or single crystal
Correction(s)	Standard Becker and Coppens extinction Absorption correction for sphere and cylinder Any user defined correction
Structure	Crystallographic (Nuclear) and/or magnetic structure
Modulation	Any number of wave vectors for magnetic moments and/or atomic population One vector only for atomic position modulation (charge density wave)

Special features	Any user defined constraints (linear or not)
	Wide range of application as (example) :
	    Form Factor fitting,
	    Anharmonic thermal factor fitting,
	    Dynamic weights (including likely hood fitting),
	    ...

REFERENCES

[1]   P. Wolfers, M. Bacman, E.F. Bertaut.
        Journal de Physique, **32**, C1–859, 1971.

[2]   M. Bacmann, P. Wolfers, G. Courbion
        Materials Research Bulletin, **15**, 1479–1487, 1981.

[3]   Bordet P., Hodeau J.L., Wolfers P., Miraglia S., Benoit A., Marezio M.,
        Remeika J.P.(1986). *Physica,* **136B**, 432–435.

[4]   Bordet P., Hodeau J.L., Wolfers P., Raggazoni J.L., Génicon J.L.,
        Tournier R., Chaudouet P., Weiss F., Espinosa G., and Marezio M. (1988).
        J. de Physique, Colloque C8, Suppl. N°12, **49**.

[5]   Collomb A., Abdelkader O., Wolfers P., Guitel J.C., Samaras D.
        J. Magn, Magn. Mat., **58**, 247-253,(1986)

[6]   Hodeau J.L., Bordet P., Wolfers P., Marezio M., Remeika J.P.
        J. Magn. Magn. Mat., **54–57**, 1527–1528.

[7]   E. F. Bertaut. *Actat Cryst.,* **24**, 217 (1968).

[8]   Barr J.R. and Heidelbrecht B.(1981). Dep. of comp. science,
        NBS PASCAL, Decus software PDP11/RSX11M Chicago mag-tape.
        University of Montana, Missoula, Montana 59812.

[9]   C.B. Shoemaker, D.P. Shoemaker, R. Fruchart.
        Acta Cryst., **C40**, 1665 (1984).

[10]   D. Givord, H.S. Li, J.M. Moreau, *Solid State Commun.,* **50**, 497 (1984) .

[11]   P. Becker, P. Coppens, *Acta Cryst,* **A30**, 129 (1974).

[12]   P. Wolfers, S. Miraglia, D. Fruchart, S. Hirosawa, M. Sagawa,
        J. Bartolome, J. Pannetier. *J. of Less Comm. Met,* To be published.

[13]   A. Fujita, H. Onodera, H. Yamauchi, M. Yamada, H. Yamamoto,
        S. Hirosawa, H. Sagawa, *J. Magn. Magn. Mat.,* **69**, 267 (1987).

Inst. Phys. Conf. Ser. No 107: Chapter 2
Paper presented at Neutron Scatt. Data Anal. Conference, Rutherford Appleton, 1990

135

On the normalization of spallation source powder diffraction data

R. O. Piltz

Department of Physics, University of Edinburgh, Mayfield Road, Edinburgh EH9 3JZ

Abstract: The normalization and absorption correction of powder diffraction data obtained from the HRPD instrument at ISIS is considered. Accuracy tests of an experimentally determined absorption correction for hydrogenous and other absorbing materials are presented. An intensity normalization procedure using diffraction powders is also presented.

1   Introduction

Spallation source powder diffraction offers the possibility of very high peak resolution and thus the ability to extract a large amount of structural information. However, the analysis of spallation source diffraction data is significantly complicated by the use of a poly-chromatic beam. Not only are extinction, sample and environment absorption, and detector efficiencies all wavelength dependent, but the intensity versus wavelength distribution of the beam striking the sample is non-trivial and must be determined experimentally.

In this work we illustrate the normalization and absorption correction of diffraction data from the HRPD instrument at the ISIS spallation source. Particular emphasis has been given to determining accurate intensities at short d-spacings, and in performing accurate absorption corrections for hydrogenous samples. Accuracy checks on the absorption correction are given and normalization procedures based on powder diffraction samples are presented. Finally an example is given illustrating the effects that inaccurate corrections have on the refined structural parameters.

2   Normalization

A schematic of the HRPD instrument is shown in Figure 1. The normalized intensity $I_{\text{norm}}(\lambda)$ and the efficiency and monitor corrected intensity $I_{\text{corr}}(\lambda)$ are given by

$$
\begin{aligned}
I_{\text{norm}}(\lambda) &= \frac{I^\star_{\text{MD}}(\lambda)/E_{\text{MD}}(\lambda)}{I^\star_{\text{US}}(\lambda)/E_{\text{US}}(\lambda)} \, A(\lambda) \, N(\lambda) \\
&= I_{\text{corr}}(\lambda) \, A(\lambda) \, N(\lambda)
\end{aligned}
\tag{1}
$$

where $I_X^*(\lambda)$ & $E_X(\lambda)$ are the focussed intensities and estimated efficiencies for the upstream monitor (US) and the main detector bank (MD), $N(\lambda)$ & $A(\lambda)$ are the normalization and sample absorption corrections factors respectively. For convenience we have expressed the time of flight intensity spectra as functions of wavelength, this however is not strictly valid for focussed intensities. For example, the main detector bank of HRPD consists of 20 concentric rings at different values of 2θ. To ensure no unnecessary degradation of peak resolution occurs, the spectra of the 20 rings are 'focussed' together, that is they are summed as functions of $d = \lambda/(2\sin\theta)$. The focussed spectrum can then be expressed in terms of the

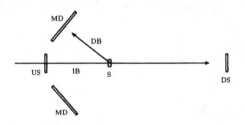

Figure 1: A schematic of the HRPD instrument at ISIS. Shown are the incident beam (IB), the diffracted beam (DB), main detector bank (MD) consisting of 20 concentric rings of detectors, the sample (S) and the upstream (US) and downstream (DS) monitors.

nominal wavelength $\lambda = 2d\sin\theta_{10}$, where θ_{10} is the Bragg angle for the 10th detector ring. To compensate for the effects of focussing, a convolution is applied to the upstream monitor spectra to emulate the focussing performed on the main detector spectra.

The normalization factor, $N(\lambda)$, includes the effects of the wavelength distribution of the beam striking the sample differing from the wavelength distribution measured by the upstream monitor; any absorption of the incident and diffracted beams by the sample environment (i.e. cryostat or furnace) or possibly the upstream monitor, the inability of the convolution applied to the upstream monitor spectrum to exactly emulate the focussing of the main detector spectra; the difference between the true and the estimated efficiencies $E(\lambda)$. In theory, the latter two effects can be reduced to negligible amounts by correctly characterising the relative detector efficiencies and the effects of focussing. The normalization factor can be determined by measuring the intensity I_{corr} for some standard sample. Two distinct types of standard samples can be used: (i) an incoherent scattering sample (typically vanadium) where I_{corr} after being stripped of any diffraction peaks is compared to the calculated incoherent scatter from the sample; (ii) a diffraction sample (such as MgO, KCl, etc.) where only the peak intensities of I_{corr} are compared to the calculated diffraction intensities.

It should be noted that the absorption of thermal neutrons by vanadium ^{51}V is accompanied by an emission of prompt ($\ll 1\mu\text{sec}$) γ-rays and by delayed γ-rays following the β-decay to ^{52}Cr (half-life$\simeq$4 min.). As the main detector bank of HRPD are γ-ray sensitive vanadium is probably unsuitable as a standard sample for this instrument and therefore diffraction calibrants were used.

The method used for calculating $N(\lambda)$ with diffraction calibrants can be summarized as follows:

1. Determine $I_{\text{corr}}(\lambda)A(\lambda)$

2. Subdivide the intensity spectrum into 0.25Å wide regions in λ

3. Perform Rietvelt refinements on each 0.25Å wide region, keeping all structural parameters fixed to the known values but refining the scale factor, background parameters, peak shape parameters, etc.

4. Use the reciprocal of the refined scale factor for each region as the value of $N(\lambda)$ for λ at the center of that region

Though this method is not an elegant one, it does have the advantage of not requiring modification of existing software. An alternative method, though not attempted, would be to replace steps 2–4 by the refinement within the Rietvelt program of a parameterised form of $N(\lambda)$ (i.e. a polynomial series or a cubic spline).

3 Absorption Correction

Two types of sample cans were used for the HRPD experiments (i) a slab can, rectangular shaped with thin vanadium windows on the front and back and four supporting walls of thick aluminium, (ii) a cylindrical 'Harwell'-type can made of thin walled vanadium. A gadolinium mask was placed in front of the slab can to shield the aluminium supporting walls from the incident beam.

For a slab can of depth D, the sample absorption factor is given by

$$A(\lambda) = \frac{\gamma \, \mu(\lambda) \, D}{1 - e^{-\gamma \, \mu(\lambda) \, D}} \tag{2}$$

The inverse absorption length, $\mu(\lambda)$, is related to the total cross-section σ_T by $\mu = N\sigma_T$ where N is the number of atoms per unit volume. The geometric factor $\gamma = (1 + |\cos 2\theta|)$ is approximated to 2, which is a valid approximation for HRPD since the main detector banks are located at $2\theta = 160$ to $178°$. The corresponding transmission factor (the attenuation of the incident beam due to sample absorption) is given by

$$T(\lambda) = e^{-\mu(\lambda)D} \tag{3}$$

For a cylindrical can of diameter D (assuming $2\theta = 180°$) we obtain

$$A(\lambda) = \frac{\pi}{2} \frac{\mu(\lambda)D}{1 - C(2\mu(\lambda)D)} \tag{4}$$

$$T(\lambda) = C(\mu(\lambda)D) \tag{5}$$

where

$$C(x) = \int_0^{\pi/2} d\alpha \, \cos\alpha \, e^{-x\cos\alpha} \tag{6}$$

Using the above formulae it is possible to calculate $A(\lambda)$ from $T(\lambda)$, the latter being obtained using the measured upstream and downstream monitor intensities.

$$T(\lambda) = \frac{\{I_{DS}(\lambda)/I_{US}(\lambda)\}_{\text{sample}} - \{I_{DS}(\lambda)/I_{US}(\lambda)\}_{\text{absorb}}}{\{I_{DS}(\lambda)/I_{US}(\lambda)\}_{\text{empty}} - \{I_{DS}(\lambda)/I_{US}(\lambda)\}_{\text{absorb}}} \tag{7}$$

where $I_X(\lambda)$ is the relevant upstream or downstream monitor intensity measured with the sample in the sample can, an empty sample can, and a sample can filled with a strong absorber such as B_4C. For the slab can arrangement the strong absorber measurement need not be performed as the downstream monitor intensity is always zero due to the presence of the gadolinium mask.

When working with slab cans a further correction is sometimes necessary due to the presence of the gadolinium mask. For a sample where μ varies with wavelength the effective center of the sample will move toward the front of the can as μ increases, resulting in less interception of the diffracted beam by the gadolinium mask. To correct for this effect the intensity is divided by

$$1 - \frac{2|\tan 2\theta|}{\pi} \left(\frac{D}{H} + \frac{D}{V} \right) \left(\frac{d}{D} + \frac{1}{2\mu D} - \frac{1}{e^{2\mu D} - 1} \right) \tag{8}$$

where H & V are the sizes of the rectangular hole in the mask and d is the distance from the mask to the sample can. The above correction is generally of the order of a few percent or less for a moderately absorbing sample unless H or V is much less than D.

4   Software Routines

The Fortran program FOCCOR calculates $I_{corr}(\lambda)$ from the raw intensity data using Equation 1. The program is a modification of the detector bank focussing program FFOCUS written by A.C.Barnes of ISIS. The intensities $I^\star_{MD}(\lambda)$ & $I_{corr}(\lambda)$ for a KCl sample are shown in Figure 2. The focussed intensities $I^\star_{MD}(\lambda)$ at both ends of the spectra are slightly greater than zero. Therefore, to ensure that $I_{corr}(\lambda)$ does not tend to infinity at the ends of the spectra, FOCCOR calculates and removes this 'background' intensity prior to division by $I^\star_{US}$. Division by the focussed $I^\star_{US}$, instead of the raw I_{US}, results in a significant improvement particularly for obtaining accurate intensities near the extremes of the spectrum. This is important when an experiment is performed which requires the shortest of d-spacing information.

The Fortran program TRANS is used to calculate the transmission factor $T(\lambda)$ using Equation 7. Early versions of TRANS used the direct division of the intensity spectra, however the estimated errors in the resultant $\mu(\lambda)D$ were poor. A significant improvement was obtained by smoothing all spectra prior to, and after, division. The prior smoothing of the I_{US} not only reduces the estimated errors in $\mu(\lambda)D$ but it also reduces the positive bias which can occur by the division process when the relative errors of the divisor are large. It should be noted that the focussing of $I^\star_{US}$, as performed by FOCCOR, also reduces this type of bias.

The GENIE command routines CYLABS & SLABABS are used to calculate $A(\lambda)$ from $T(\lambda)$ using Equations 2–6, and then to multiply $I_{corr}(\lambda)$ by $A(\lambda)$. The user selects between $\mu(\lambda)D$ obtained directly from $T(\lambda)$, or a weighted linear fit in λ can be applied to $\mu(\lambda)D$. The option also exists for an alternative linear relationship between $\mu(\lambda)D$ and λ to be given. For the majority of samples it is expected that the linear fit option will be most appropriate. However for our hydrogenous samples a linear relationship was found to be inadequate in describing the variation in $\mu(\lambda)D$. The function

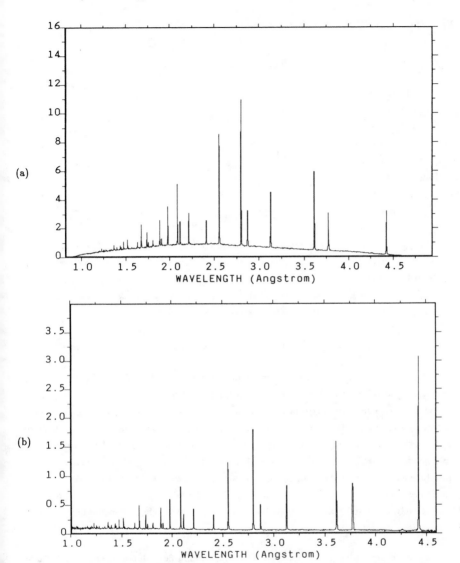

Figure 2: Intensity spectra obtained from a KCl sample using the HRPD instrument at ISIS; (a) the focussed main detector intensity $I^\star_{MD}(\lambda)$; (b) the efficiency and monitor corrected intensity $I_{corr}(\lambda)$. The displayed intensity scales are arbitrary.

$C(x)$ used in the cylindrical sample absorption correction was calculated numerically, tabulated, and a linear interpolation method used to approximate its value and that of its inverse.

Finally, the correction of $I_{corr}(\lambda)$ for the normalization factor is performed using the GENIE routine NORCOR which linearly interpolates $N(\lambda)$ from a given table of values.

5   Absorption Corrections Tests

In the first test the calculated and measured μ for
vanadium metal are com-
pared. Four plates of vanadium metal with a com-
bined thickness of 14.5(5)mm
were used for the measurements. The calculation of
μ is based on the literature
value for the vanadium's density, 6.11g/cc, and the total cross-section of vanadium
as a function of wavelength
given by Granada, Kropff &
Mayer(1981). The calculated
and observed μD (Figure 3)
agree within 3% over the ma-

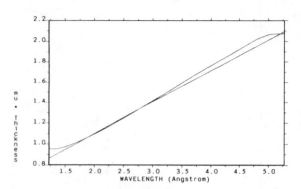

Figure 3: The measured (curved line) and calculated (almost straight line) values of μD versus λ for a 14.5mm thick vanadium plate. The tendency of the measured μD to have a zero gradient at either side of the spectra is an artifact due to the smoothing of the monitor spectra.

jority of the spectrum indicating the possibility of very accurate absorption corrections.

The second test used 5 & 15mm thick slab cans containing weighed quantities of KH_2PO_4. The sample absorption was considerable for both volumes of sample; for the 15mm thick sample μD varied from 0.7 to 1.7 over the measured wavelength range of 0.5 to 4.5Å. As the effective absorption of KH_2PO_4 is not known we were unable to compare the measured μD with a known wavelength dependence. However,the absorption corrected peak intensities agreed very well, when corrected for the ratios of the sample weights, as seen in Figure 4. A direct comparison of the two spectra is complicated by the slightly broader peaks from the 15mm sample, however if the peak areas are compared then the agreement is even better. For the peaks near $\lambda=3.8$ & 3.9Å the difference in peak areas is only 2.4 & 1.3% respectively, the absorption corrections for both of these peaks being approximately 1.6 & 3.5 for the 5 & 15mm thick samples.

The third test on absorption used a sample of CrOOH in a 15mm deep slab can, and in a 11.7mm diameter cylindrical can. The absorption corrected intensities were a factor of approximately 1.49 larger in the slab can, which is realistic in terms of the relative amounts of sample in the incident beam. The absorption corrections ranged from 1.2 to 2.8 for the cylindrical can and from 1.6 to 3.4 for the slab can. Comparing the eight largest peaks, the ratio of peak areas from the two samples 1.49 to within $\pm 5\%$. The absorption corrections for these peaks ranged from 1.8 to 2.6 for the cylindrical sample and from 2.3 to 3.3 for the slab can sample. The relatively poor agreement between peak areas may be due to the inhomogeneous character of the beam profile. As the slab and cylindrical cans intercept different portions of the beam, which is known not to have a constant wavelength distribution across its profile, then the averaged wavelength distribution seen by both samples will differ. It should be noted that beam homogeneity is important for the absorption correction of a cylinder. For instance, a greater beam intensity exists near the axis of the cylinder, as compared to the edges, then the sumptions leading to Equations 4 & 5 will be invalid, and these equations will calculate $A(\lambda)$ & (λ) to be too small and too large respectively. Fortunately, the effects of beam homogeneity will be partially cancelled in the calculation of $A(\lambda)$ from $T(\lambda)$.

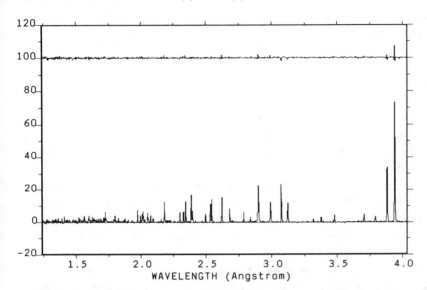

Figure 4: The absorption corrected intensity $I_{corr}(\lambda)$ $A(\lambda)$ divided by sample mass for a CrOOH sample in a 5mm thick slab sample can. Shown on the same scale but shifted by 100 intensity units is the difference between the absorption corrected intensity divided by sample mass for a 5mm and a 15mm thick sample. To aid visual comparison the spectra backgrounds were removed and the spectra divided by λ.

Normalization using Diffraction Calibrants

ll measurements were performed using 11.7mm cylindrical sample cans. The diffraction calibrants sed were MgO & KCl, no sample environment equipment was used with the MgO sample, whereas the Cl sample was placed inside a high temperature furnace. Measurements were also made of powdered anadium samples both inside the furnace and without the furnace.

sing an optical microscope the grain sizes of the KCl and MgO powders were measured to be 1–3 & –5μm respectively. For such a grain size the extinction effects in KCl will be negligible (<1%) for the avelengths used. The MgO grains observed under the microscope were irregular shaped and showed onsiderable internal fracturing indicating the effective domain size is probably much less than the 3– μm. This is consistent with the particle size broadening observed for MgO being significantlye small (<1.5%). T

IgO based on a domain size of 3–5μm would reduce $N(\lambda)$ by $\simeq 25\%$ at $\lambda = 4$Å and $\simeq 6\%$ at $\lambda = 2$Å.

The structural parameters for MgO & KCl at room temperature were taken from Barron(1977) and Cooper & Rouse(1973) respectively. The resultant normalization factors $N(\lambda)$, assuming no extinction effects, are shown in Figure 5.

For a sample not in the furnace $N(\lambda)$ is roughly constant at large λ, with an approximate 20% increase in $N(\lambda)$ as λ approaches 0.5Å. This appears to indicate that the efficiency of the main detector bank is being overestimated at short wavelengths. Comparing $N(\lambda)$ for the sample in the furnace with $N(\lambda)$ not in the furnace we find that the absorption of neutrons by the furnace walls is considerable and increasing with increasing wavelength. It is unclear if the sharp increase in $N(\lambda)$ determined using KCl for $\lambda > 4.2$Å

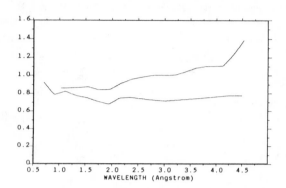

Figure 5: The normalization factors $N(\lambda)$ determined using KCl within a furnace (upper spline), and MgO not in the furnace (lower spline). Both $N(\lambda)$ have been multiplied by constant factors to bring the values close to unity.

is due to the furnace absorption or some other effect. One possibility is that the convolution of $I_{US}(\lambda)$ to simulate focussing is inaccurate. (The range in wavelength of the incident neutrons was different for the furnace and non-furnace measurements.) The common dip in $N(\lambda)$ for $\lambda \simeq 1.9$Å results from the beam being effectively narrower at the sample position for this particular wavelength. This beam narrowing can also be observed in the intensity of the downstream monitor when a strongly absorbing sample is placed in the sample can.

Of the above features of $N(\lambda)$, the beam attenuation due to the furnace and the error in the estimation of the detector efficiency should remain constant with time and be independent of the sample can used. On the other hand the beam narrowing effect is dependent on the size and geometry of the sample can and therefore a normalization run is needed for each type of sample can used. It is also expected that the beam narrowing effect will vary with time due to changes in the beam profile.

7   Normalization using Vanadium

When using vanadium as a normalization calibrant the incoherent scatter from the sample can and the sample environment must be removed. This was performed using the measured transmission factor of the vanadium sample, and the $I_{corr}(\lambda)$ for an empty can and a sample can containing a strong absorber (Lindley & Mayers, 1988). The resultant intensity was then corrected for absorption using the TRANS & CYLABS routines. The multiple scattering correction was determined using a Monte-Carlo technique assuming isotropic and elastic scattering. The multiple scattering contribution to the intensity varied between 6 & 7%. The Placzek variation of the differential cross-section of vanadium with wavelength was compensated for by using the values of Mayers(1984). Finally, the normalization factor $N(\lambda)$ is obtained by taking the reciprocal of the final vanadium spectra after any diffraction peaks have been removed from that spectra.

The ratios of the $N(\lambda)$ determined using diffraction and vanadium calibrants are shown in Figure 6. Ideally the ratios should be constant with respect to wavelength, which is not the case. It is not possible to determine how much of this discrepancy is due to the γ-ray sensitivity of the main detectors as there appears to be contaminating material in the vanadium powder. The measured $\mu(\lambda)$ D for both vanadium samples is approximately 50% larger than the calculated values using the weighed mass and volume of the vanadium powder in the sam-

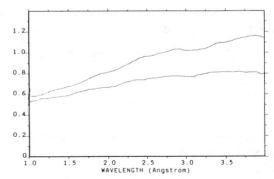

Figure 6: $N(\lambda)$ determined using KCl or MgO divided by $N(\lambda)$ determined using vanadium. The upper plot is for measurements within the furnace, the lower plot for measurements without the furnace. Both ratios have been multiplied by constants so that they fit on the same scale.

ple can. It is therefore possible that the contaminant (possibly hydrogen chemically bonded to the vanadium) is affecting the measured vanadium intensity. However, the presence of a contaminant material cannot explain why the final corrected intensity from the vanadium in the furnace increases at longer wavelengths relative to the non-furnace vanadium intensity. This effect can be explained if the absorption for the prompt γ-rays through the furnace walls is low, in such a case the proportion of γ-rays detected to neutrons detected would increase with wavelength.

8   Refinements

The compound $D_2C_4O_4$ was measured in the furnace at room temperature and intensities were corrected using $N(\lambda)$ determined by both the KCl calibrant and the vanadium powder calibrant. The positional and anisotropic temperature factor parameters for all the atoms were fixed to the values of a single crystal neutron diffraction study (McMahon et al., 1990) and only the extinction, preferred orientation, scale, background, and peak shape parameters were refined. The integrated intensity R-factor for the MgO calibrated data was significantly better 10% compared to 15% for the vanadium calibrated data. (The poor R-factors are a result of poor counting statistics due to a quick measurement.)

During the refinement of $D_2C_4O_4$ data collected at a temperature of 240C, the effect of three different $N(\lambda)$ values was investigated. For the first the values obtained using the KCl calibrant were used, the second also used the KCl calibrant values but with a dip of 10% in the region between 1.5 &

1.7Å, for the third $N(\lambda)$ was set to unity. Full Reitveld refinements on all positional and anisotropic temperature parameters were performed on the three differently normalized data sets. As expected the third data set gave significantly different temperature parameters compared to the first data set, however the positional parameters agreed to within 2 esd's. The refinement of the second data set also produced different temperature factors but as well the positional parameters were considerably different compared to the first data set. In particular the z value of the the D atom moved by 5 esd's corresponding to a shift of 0.02Å.

9 Summary

The sample absorption corrections routines used give highly accurate results for the case of slab sample cans. Some doubt remains as to the actual accuracy obtainable using cylindrical sample cans, though it appears to be better than 5%.

Despite the fact that vanadium is well understood as an incoherent scattering standard, its applicability in the case of γ-ray sensitive detectors cannot be assumed. Other possible incoherent scattering materials that do not emit γ-rays may however be applicable. The use of diffraction samples for the determination of the normalization factor appears promising, though further work is required to determine the accuracy obtainable with such procedures. One problem with the use of diffraction calibrants is the need to have samples that are known to be free from extinction effects. As well, diffraction calibrants only provide the values of $N(\lambda)$ at discrete values of λ corresponding to peak positions. In practice more than one diffraction calibrant should be used to enable a check on accuracy and to ensure that all regions in λ are covered.

Finally, test refinements indicate that inaccurate normalization corrections as small as 10% result not only in inaccurate temperature parameters but can produce significantly incorrect positional parameters.

References

Barron, T.H.K 1977, *Acta Cryst.* **A33**, 602–4.

Cooper, M.J. and Rouse, K.D., 1973, *Acta Cryst.* **A29**, 514–20.

Granada, J.R., Kropff, F. and Mayer, R.E., 1981, *Nucl. Instr. and Meth.* **189**, 555-9.

Lindley, E.J. and Mayers, J. 1988, in *Neutron Scattering at a Pulsed Source*, R.J. Newport (ed), Institute of Physics Publishing.

McMahon, M.I., Kuhs, W.F., Nelmes, R.J. and Semmingsen, D., 1990, *in preparation*.

Mayers, J. 1984, *Nucl. Instr. and Meth.* **221**, 609–18.

Inst. Phys. Conf. Ser. No 107: Chapter 2
Paper presented at Neutron Scatt. Data Anal. Conference, Rutherford Appleton, 1990

The data analysis of reciprocal space volumes

C C Wilson

ISIS Facility, Rutherford Appleton Laboratory, Chilton, Didcot, Oxon OX11 0QX, UK.

ABSTRACT: The ISIS Single Crystal Diffractometer SXD is a time–of–flight Laue instrument exploiting the pulsed nature of the ISIS beam along with large area detectors. This means that the data collected on SXD is in the form of reciprocal space volumes. The data analysis procedures used on SXD are described and illustrated. Examples of some of the novel applications of the instrument are given.

1. INTRODUCTION

Single crystal diffraction at a pulsed neutron source is an area rich in information. By the utilisation of both the time–of–flight technique to sort the white beam and large area position–sensitive detectors, it is possible to access large volumes of reciprocal space in a single measurement. The ISIS Single Crystal Diffractometer SXD (Figure 1; Forsyth and Wilson, 1990) is just such a time–of–flight Laue instrument. The instrument parameters are given in Table 1. The main features of the instrument are evident from Figure 1 and these parameters. The scintillator–based position–sensitive detectors referred to are of the Anger camera design (Anger, 1958; Forsyth, Lawrence and Wilson, 1988) whose performance on SXD is discussed elsewhere (Wilson, 1989; Wilson and Zaleski, 1990). It is the recent successful commissioning of these large area detectors (Wilson and Zaleski, 1990) which has enabled the SXD scientific programme to begin in earnest, as discussed below.

The ability of such an instrument to probe reciprocal space volumes is illustrated in Figure 2. The combination of a time–sorted wavelength range (providing two Ewald sphere surfaces at $1/\lambda_{max}$ and $1/\lambda_{min}$) and the large scattering locus subtended by the detector, allows the area shown in Figure 2 to be accessed *in a single measurement*. Of course, the detector also has height, and a full 3–dimensional reciprocal space volume is accessed on SXD.

The simplest way of initially visualising SXD data is in a time–condensed picture, showing the detector surface. This is of course entirely analagous to any white beam Laue plot, once the time–of–flight discrimination is removed in this way. A typical Laue plot from SXD is shown in Figure 3.

Figure 1 – Plan view of SXD. Note the provision of 3 large area PSDs subtending a large scattering angle at the sample.

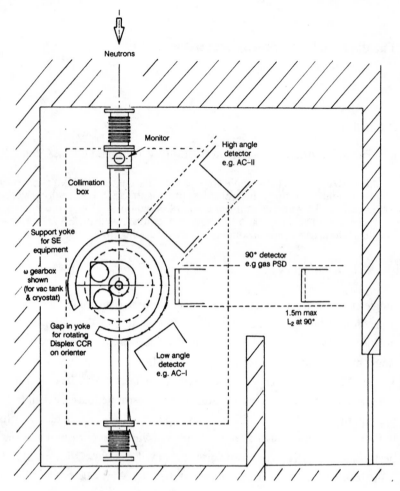

TABLE 1 – SXD instrument parameters

Laue time–of–flight diffractometer
Primary flight path $L_1 = 8m$ from 316K H_2O moderator
Sample–detector distance $L_2 = 0.2–1.5m$ (longest at 90°)
Wavelength range 0.25–8Å
$\sin\theta/\lambda$ range $0.02–2.5\text{Å}^{-1}$

Detectors
3 x large area Position–sensitive detectors

> Anger camera I, ZnS scintillator, >300 x $300mm^2$, 7–8mm resolution
> Anger camera II, ZnS scitnillator, >350 x $350mm^2$, 5mm resolution
> ORDELA ^{3}He gas detector, 250 x 250mm, 2–8mm resolution

Resolution, $\Delta d/d = 10^{-2} – 5.10^{-3}$

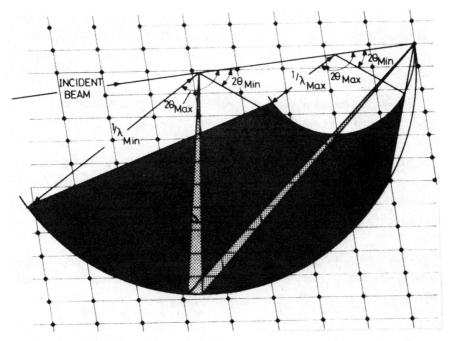

Figure 2 – Single crystal time–of–flight geometry.

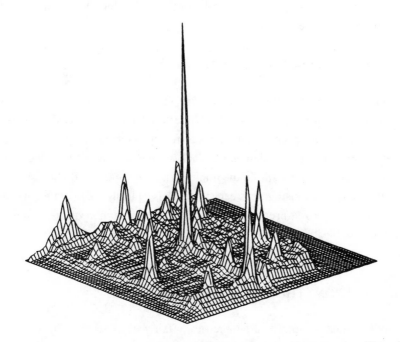

Figure 3 – Laue plot of intensity from SrF_2 measured on the SXD Anger camera. The reflections contributing to the strong central peak are the (00l) row. This was in fact the first successful Laue pattern measured with an area detector on SXD.

It is clear from the above that SXD is an ideal instrument for the surveying of reciprocal space. This capability has particular relevance in the study of incommensurate structures, phase transitions and diffuse scattering. In these fields the time–of–flight Laue technique is extremely powerful.

However, the characteristics of the data collected on such an instrument can also have certain advantages in more routine structural work :

(i)   The collection of many Bragg reflections simultaneously in the detector allows the accurate determination of crystal cell and orientation from a single data histogram (collected in one fixed crystal/detector geometry). It should also be noted that for some applications this single histogram may be the only data required;

(ii) The white nature of the incident beam enables the straightforward measurement of reflections at different wavelengths. This ability is invaluable in the precise study of wavelength dependent effects such as extinction and absorption;

(iii) The collection of data to the very high $\sin\theta/\lambda$ values accessible on SXD (exploiting the high flux of useful epithermal neutrons from the undermoderated ISIS beams) allows more precise parameters to be obtained, enabling the examination of very subtle structural features.

2. THE PROBLEM : VOLUMES OF DATA

The trouble with large volumes of reciprocal space is that in examining them, one obtains large volumes of diffraction information (scattered intensity measurements).   SXD typically operates with several thousand detector elements (a single Anger camera has 64x64 pixels = 4096 detector elements), which when combined with a few hundred time channels, essential of course within the framework of the time–of–flight technique, tends to give rather large data histograms, each of which is typically accumulated in less than an hour.   Hence a major requirement for the efficient exploitation of the technique is the availability of a flexible software package for data analysis and display (Forsyth et al, 1986).

3. THE SOLUTION : THE SXD SOFTWARE PACKAGE

Fortunately, crystallography is a geometrically fairly exact science, in which from the initial provision or extraction of a little information regarding the crystal, most of the data analysis procedures can be made routine and almost automatic.   This is illustrated in equation (1), the equation governing data collected on a 4–circle diffractometer, given by Busing and Levy (1967),

$$h_L = \Omega \, \chi \, \Phi \, \mathbf{U} \, \mathbf{B} \, h \tag{1}$$

where

h_L is the diffraction vector in the laboratory Cartesian system;

Ω, χ and Φ are the matrices defining the diffractometer settings;

$\mathbf{U}$ and $\mathbf{B}$ are matrices describing the sample;

h is the reciprocal lattice vector.

SXD is designed, as far as possible, in accordance with standard 4–circle geometry, and is probably best defined as a "restricted 4–circle" diffractometer.

The terms Ω, χ and Φ are defined by the diffractometer setting angles and are thus known. The h_L vector (peak coordinates in the laboratory Cartesian system) is measured on the detector. The crystal parameters are taken into account in the $\mathbf{U}$ and $\mathbf{B}$ matrices. The $\mathbf{B}$ matrix in equation (1) is the reciprocal metric tensor, defined as

$$\mathbf{B} = \begin{bmatrix} b_1 & b_2\cos\beta_3 & b_3\cos\beta_2 \\ 0 & b_2\sin\beta_3 & -b_3\sin\beta_2\cos\alpha_1 \\ 0 & 0 & 1/a_3 \end{bmatrix} \tag{2}$$

where the a_i, α_i and b_i, β_i refer to the real and reciprocal cell, respectively, and is thus determined by the unit cell of the sample. The $\mathbf{U}$ matrix is the orthogonal matrix which defines the crystal orientation with respect to the diffractometer coordinate system. The $\mathbf{U}$ and $\mathbf{B}$ matrices are normally combined into a single UB matrix, sometimes loosely referred to as the orientation matrix. The determination of the UB matrix is the final stage in the definition of terms in equation (1). The aim of the early stages of data analysis is thus to define the sample and diffractometer–dependent terms in equation (1), to allow its application, for example, in the form for peak indexing

$$h = \mathbf{B}^{-1} \, \mathbf{U}^{-1} \, \Phi^{-1} \, \chi^{-1} \, \Omega^{-1} \, h_L \tag{3}$$

The determination of the UB matrix and its use in peak indexing is normally one of the first steps in an SXD experiment. The central role of this procedure in the SXD software is shown in Figure 4.

The dual purpose of SXD is evident in the portion of this Figure beyond UB matrix refinement. The left–hand path "Structural Refinement" is routine crystallographic work, involving the extraction of structure factors and subsequent least squares refinement, while the right–hand side "Reciprocal Space Surveying" attempts to exploit the special ability of a time–of–flight Laue diffractometer such as SXD to sample large reciprocal space volumes.

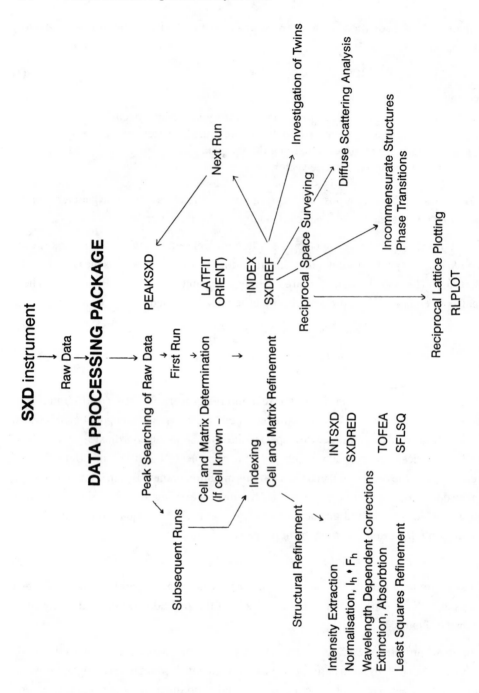

Figure 4 – The SXD data processing package

4. STRUCTURAL REFINEMENT

The characteristics of the structure factor data collected on an instrument such as SXD have certain advantages in structural refinement, outlined in the Introduction above. The extraction of structure factor information from an SXD data set proceeds in two stages. Given the provision of an accurate UB matrix from the earlier stages of the software package, integrated intensities are extracted from the raw data using Wilkinson's modification of the $\sigma(I)/I$ method (Wilkinson, 1986; Wilkinson et al, 1988). This procedure involves fitting variable–shaped ellipsoidal integration volumes around the reflection position as predicted from the UB matrix by equation (1). Local interpolation of the peak centroid improves the integration accuracy and can also be used to improve the quality of the UB matrix. The "learning" of the shapes of the stronger reflections is exploited by imposing these shapes upon the profiles of weaker reflections occurring in similar regions on the detector. The use of this peak shape library method allows significantly more reliable weak intensities to be extracted, along with the benefit of more accurate strong intensities from the application of the $\sigma(I)/I$ technique. Since reflections at high Q tend to be rather weak, the availability of data at very high $\sin\theta/\lambda$ values of $>2\text{Å}^{-1}$ on SXD depends strongly on the successful and reliable extraction of weak peak intensities.

Once a reliable set of integrated intensities has been extracted from the raw data, it is necessary to reduce these to structure factors. For this we use the formula of Buras and Gerward (1975)

$$I_h = i_0(\lambda) \ V \ N^2 \ |F_h|^2 \ \lambda_h^4 \ \epsilon(\lambda,\alpha) \ A_h(\lambda) \ E_h(\lambda) \ / \ 2\sin2\theta_h \tag{4}$$

where

> I_h is the measured intensity of reflection h;
> $i_0(\lambda)$ is the incident flux;
> V is the crystal volume;
> N is the number density of unit cells;
> $|F_h|$ is the structure factor magnitude of reflection h;
> λ_h is the wavelength at which h is measured (λ^4 is reflectivity);
> $\epsilon(\lambda,\alpha)$ is the detector efficiency, a function of wavelength and detector coordinate $[\alpha = \alpha(x,z)]$;
> θ_h is the Bragg angle at which h is measured (The term $1/2\sin2\theta_h$ is the Lorentz correction. It should be noted that in time–of–flight Laue diffraction, the "moving" part to which this correction refers is the contraction of the Ewald sphere (radius $1/\lambda$) during the pulse time frame);
> $A_h(\lambda)$ is the absorption correction;
> $E_h(\lambda)$ is the extinction correction.

In practice in the SXD package this data reduction is accomplished in three stages :

(i)   The detector spatial response correction $\epsilon[\alpha(x,z)]$ is taken account of within the peak integration program;

(ii)   In the main reduction program SXDRED, the expression evaluated is

$$|F_h| = \{I_h\, 2\sin 2\theta_h\, /(i_o(\lambda)\, V\, N2\, \lambda_h^4\, \epsilon(\lambda))\} \tag{5}$$

(iii) In the CCSL (Brown and Matthewman, 1987) routine TOFEA, the coefficients for the evaluation and subsequent refinement of wavelength and path length dependent absorption and extinction corrections are calculated.

Structure factor refinement is then carried out within the framework of CCSL using the least squares program SFLSQ.

4.1 Example – wavelength dependent extinction and anharmonic thermal vibrations in strontium fluoride (Forsyth, Wilson and Sabine, 1989).

For this experiment a cylindrical sample (height 8mm, diameter 2.5mm) of the fluorite material SrF_2 mounted along $<1\bar{1}0>$ was used. In this case data were collected solely in the equatorial plane of the instrument, allowing access to (hhl) reflections. The scattering angles for the data collection ranged from 50–110° in 2θ, and the wavelength range from 0.36–6Å. The resulting 101 reflections were used in SFLSQ to refine the structural parameters (Wilson and Forsyth, 1989). It should be noted that not all of these reflections have unique h,k,l values, since identical reflections measured at different wavelengths remain distinct observations until all wavelength dependent terms are corrected for – in this case with extinction refined in the least squares program, these reflections are used in the refinement data set. Extinction was corrected for using a variable wavelength adaptation of the CCSL extinction correction routine using a Becker–Coppens Gaussian model, with one variable parameter. The refined parameters are shown in Table 2.

TABLE 2 – Refined SrF_2 structural parameters

Scale	9.13
B_{Sr}	0.470(26) Å^2
B_F	0.732(31) Å^2
mosaic spread	1.03(10) 10^{-4} rad^{-1}
R (unweighted)	0.050

On completion of this refinement, however, it was clear that allowance for anharmonic thermal vibrations of the fluorine atoms would significantly improve the quality of the fit. To this end, a further data set was collected at $2\theta = 90°$. This data set comprised two parts :

(i) 69 (hhl) reflections measured in a full 360° scan, with all 4 symmetry equivalent reflections averaged;

(ii) 19 further reflections, again of type (hhl), collected at higher statistics and chosen since they are expected to exhibit the effects of anharmonicity in a most obvious way. These reflections were selected using the rationale given below.

It was shown by Mair and Barnea (1971) that the effect of anharmonicity of the thermal vibrations of the tetrahedrally bound fluorine atoms in SrF_2 is to give the following expression for the ratio of structure factors

$$F_+/F_- = 1 - (2b_F/b_{Sr}) \exp \{(B_{Sr}-B_F)[(h^2 + k^2 + l^2/4a^2)]\} \times$$

$$(B_F/4\pi a)^3 (\beta_F/kT) [|h_1k_1l_1| + |h_2k_2l_2|] \qquad (6)$$

where

      $|F\pm|$ are the structure factors for reflections with $|h| + |k| + |l| = 4n\pm1$;
      b_{Sr}, b_F are the scattering lengths of Sr and F;
      B_{Sr}, B_F are the isotropic temperature factors;
      a is the cubic cell parameter;
      k is the Boltzmann constant;
      T is the absolute temperature;
      β_F is the anharmonicity parameter for fluorine.

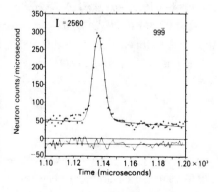 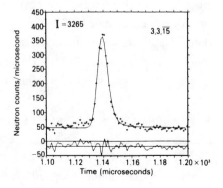

Figure 5 – the $(99\bar{9})$ and $(3,3,\overline{15})$ reflections from SrF_2, measured on SXD. In the harmonic approximation their intensities would be equal, but anharmonic vibrations of the fluorine atom makes the $4n + 1$ reflection $(3,3,\overline{15})$ more intense.

It can be seen from this that if one measures pairs of reflections with equal $(h^2 + k^2 + l^2)$, which in the harmonic approximation have equal intensities, then by (6) some intensity variation can be expected for pairs whose $|h| + |k| + |l| = 4n \pm 1$ (see Figure 5). It can also be seen from equation (6) that the effect becomes more pronounced at higher Q – this is where the ability of SXD to exploit the useful ISIS epithermal flux and access the highest resolution data becomes especially valuable. The very high $\sin\theta/\lambda$ reached in the SXD experiment (1.7Å^{-1}) allows greater precision to be obtained in the determination of the anharmonicity parameter β_F.

The anharmonicity refinements were carried out in an adaptation (Mair and Barnea, 1971) of the ORFLS program (Busing, Martin and Levy, 1962), using data pre–corrected for extinction in CCSL. The results of the refinements are shown in Table 3.

TABLE 3 – Anharmonicity refinements on SrF_2

Reflections used	Anharm Refined	Scale Factor	B_{Sr} (Å^2)	B_F (Å^2)	β_F $(\times 10^{-19} \text{ J.Å}^{-3})$	R
ALL (88)	NO	1.006(8)	0.454(11)	0.737(16)		0.0425
ALL (88)	YES	1.007(6)	0.457(10)	0.737	–5.0(1.5)	0.0388
$4n \pm 1$ (16)	NO	1.100(41)	0.539(30)	0.732		0.0441
$4n \pm 1$ (16)	YES	1.107(9)	0.539	0.732	–4.02(28)	0.0219

The improvements in fit are apparent in both cases, but are of course especially emphasised when only those pairs of reflections exhibiting anharmonicity are used.

The β_F parameter obtained from the refinement should be compared with that obtained from a least squares fit of the observed F_+/F_- ratios $(-4.19(30) \times 10^{-19} \text{ J.Å}^{-3})$ and the previous determination by Cooper and Rouse (1971) $(-3.95(46) \times 10^{-19} \text{ J.Å}^{-3})$ from reactor data. These determinations of the β_F parameter agree well with each other, but it is plain that the *precision* of the SXD determination is higher, reflecting the high $\sin\theta/\lambda$ reached (1.7Å^{-1}) compared with that attained in the earlier study (0.9Å^{-1}).

5. RECIPROCAL SPACE SURVEYING

The other aspect of SXD, indicated on the right hand side of Figure 4, is to fully exploit the reciprocal space volumes accessed by the instrument in a single measurement. The ability to probe routinely those regions of reciprocal space *between* Bragg peaks lends the time–of–flight Laue technique great power in the study of incommensurate structures, phase transitions and diffuse scattering. Examples of the use of SXD in these cases are discussed below.

The simplicity of reciprocal space surveying on SXD again arises from the straightforward application of equation (1). In this case, one chooses a Q region (i.e. a set of indices $h_{min} < h < h_{max}$, $k_{min} < k < k_{max}$, $l_{min} < l < l_{max}$) within which one wishes to probe, with application of the equation as written giving the necessary detector coordinates to allow the data to be accessed. In the SXD package this is accomplished within the RLPLOT routine. Of course a mapping of reciprocal space points onto detector coordinates will never be exact and interpolation is adopted within the extraction program. Increased flexibility is also available within RLPLOT from the optional ability for performing detector and/or incident flux normalisation.

5.1 Example – W–Hexaferrite reciprocal lattice plots (Moze and Wilson, 1990)

The W–type hexaferrite $Ba(Fe,Co)_2Fe_{16}O_{27}$ has a hexagonal unit cell (space group $P6_3/mmc$) of some $5.94 \times 5.94 \times 32.9$ Å. With such a long c axis, the (00l) reflections ought to appear in time–of–flight as a rich Laue row, with many orders observed. The crystal morphology was ideal for such a scan, having a large hexagonal (001) face. The sample was aligned in SXD with the detector centred at $2\theta = 90°$ and this crystal face at some $45°$ to the beam – thus the (00l) reflections should be found close to the centre of the detector. This material was in fact the first for which the full SXD software package (Figure 4) was used, with cell and orientation determination, indexing, cell refinement and intensity extraction proceeding routinely. The refinement of the structure is in progress. However, for the purposes of the current discussion, the UB resulting from RAFSXD was used in RLPLOT to generate the reciprocal lattice plots shown in Figure 6. This series of plots shows a range of (h0l) sections through the SXD data from the W–hexaferrite sample *from a single histogram*. As can be seen from Figure 6b, the (0,0,70) reflection is clearly visible in the raw data. It can also be seen, from the time–of–flight spectrum (Figure 7) and from Table 4 (intensities extracted by "profile" fitting the (00l) row) that useful intensity information in this material, with a >30Å cell edge, is accessed to at least the (0,0,80) reflection (d = 0.4Å, $\sin\theta/\lambda = 1.2$Å^{-1}).

TABLE 4 – Intensities from profile fit of the (00l) row of W–hexaferrite

h	k	l	I	σ(I)	h	k	l	I	σ(I)	h	k	l	I	σ(I)
0	0	80	3611	1287	0	0	78	−56	617	0	0	76	2490	1018
0	0	74	−34	534	0	0	72	9466	1525	0	0	70	14803	1772
0	0	68	1019	677	0	0	66	3942	927	0	0	64	−25	426
0	0	62	1503	596	0	0	60	2542	694	0	0	58	2060	613
0	0	56	9554	981	0	0	54	33102	1607	0	0	52	14999	1057
0	0	50	6579	672	0	0	48	1679	375	0	0	46	116	202
0	0	44	2457	328	0	0	42	18351	680	0	0	40	6349	389
0	0	38	158	113	0	0	36	1632	172	0	0	34	21671	471
0	0	32	5200	212	0	0	30	7446	217	0	0	28	28877	362
0	0	26	5521	139	0	0	24	298	32	0	0	22	697	36
0	0	20	3716	66										

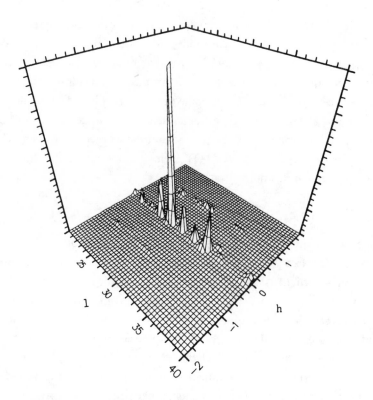

Figure 6a – The (h0l) section of data from the W–hexaferrite, in the range 20 < h < 40.

Figure 6b – A further section through the same data histogram, showing reflections in the range 60 < h < 70.

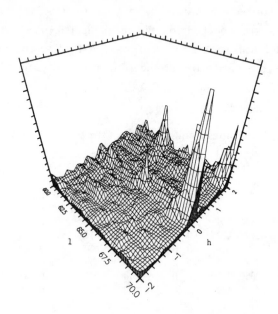

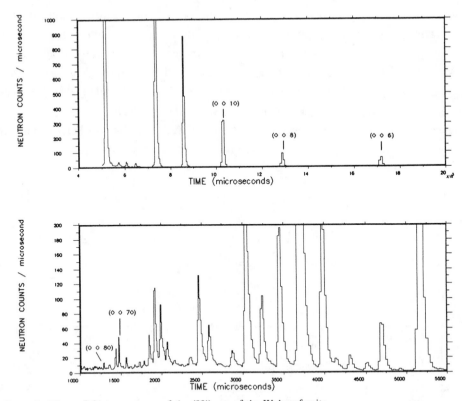

Figure 7 – Time-of-flight spectrum of the (00l) row of the W–hexaferrite.

5.2 Example – Diffuse scattering from Yttria–stabilised Zirconia (Hull et al, 1990)

Within the cubic structure of Zirconia (ZrO_2), which is stabilised at ambient temperature by the presence of $>9mol\%$ Yttria (Y_2O_3), oxygen vacancies exist to maintain electrical neutrality. While the structure of ZrO_2/Y_2O_3 is fluorite, within the crystal there occur vacancy–free regions which cause a slight tetragonal distortion of the cubic lattice. This effect will give rise to coherent diffuse peaks at systematically absent Bragg peak positions. In addition, there is some ordering of vacancy pairs within the structure, giving rise to superlattice diffuse peaks in positions away from the Bragg peaks.

Figure 8 shows data collected on SXD from a single crystal of ZrO_2–9.4%Y_2O_3. The sample was aligned with $<1\bar{1}0>$ vertical, allowing access to (hhl) reflections in the equatorial plane of the detector. The figure shows a multiplot of data collected in this equatorial plane. The 2θ values at which these data were collected range from 60° at the top to 120° at the bottom. The plot clearly shows Bragg and diffuse scattering from this material. The sharp reflections such as (113) and (224), with h,k,l all even or all odd, are allowed in the cubic space group. Others, such as (114) and (116), are the diffuse Bragg peaks due to the distortion of the cubic lattice. There is also some diffuse scattering in the regions away from the Bragg peak positions, the observation of which is made easier by the surveying capability of SXD. These superlattice peaks are those caused by vacancy clustering. The measurement of diffuse scattering to the high Q values accessible on SXD (note the observation of the (118) diffuse peak) is important for the reliable modelling of the structural disorder causing this intensity. As stated above, the plot shows results from a single row of pixels across the equatorial plane of the SXD detector. Further information is of course available simultaneously in the non–equatorial regions.

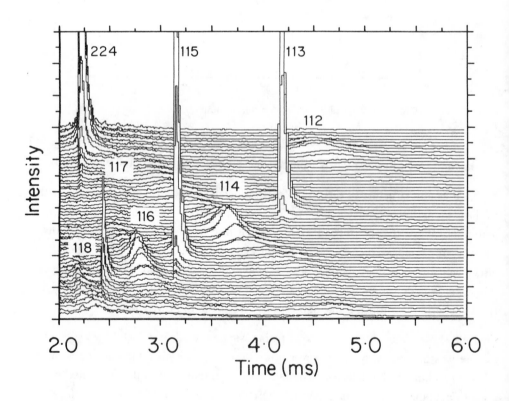

Figure 8 – Bragg and diffuse scattering in the (hhl) plane of yttria–stabilised zirconia, ZrO_2–9.4%Y_2O_3.

5.3 Example – magnetic phase transition in Holmium (Stirling et al, 1990)

The low temperature diffraction pattern of Holmium is rich in magnetic structure. The reciprocal space scanning properties of SXD can be used to follow phase transitions such as that in Ho (at around 133K) in a particularly simple way. By retaining the crystal and detector geometry fixed in one orientation it is possible to measure the temperature dependence of many reflections simultaneously. Satellite reflections, their positions and intensities, can be examined in any direction in reciprocal space. In an experiment such as the one in this example, one SXD histogram is frequently sufficient to give all the required data under one set of sample conditions.

In this experiment, the region around (00l) was examined in the temperature range 150 → 14K. The temperature variation of the (00l) row is shown in Figure 9. The magnetic phase transition gives rise to incommensurate reflections at $(0,0,l\pm q)$ which can clearly be seen to appear as the temperature is lowered from 150K to 120K. The more detailed plot of the 18K data in Figure 10 indicates the richness of the magnetic structure, with many weak peaks evident in the (00l) row. The use of the SXD software package in the study of this magnetic phase transition is illustrated in two ways:

(i)   The peak indexing of the strongest reflections in the 120K data (Table 5) shows clearly the distinctly non–integral indices of some of these reflections. The reflections in this Table show clearly the existence of an incommensurate structure with a vector $q = 0.27c$, observed around (002), (004), (015) and (006).

Table 5 – List of indexed peaks for 120K Ho data (from INDEX)

Time	Zcel	Xcel	h	k	l	lambda	RAGi(%)
10232	3.29	4.14	0.00	0.01	1.74	4.9055	26.7
8819	3.52	3.57	0.00	0.00	2.01	4.2280	1.5
7774	3.64	3.74	0.00	0.00	2.28	3.7269	28.8
4743	3.65	3.65	–0.01	–0.01	3.74	2.2738	27.3
4422	3.65	3.64	–0.01	–0.01	4.01	1.1202	2.3
4142	3.65	3.64	–0.01	–0.01	4.28	1.9859	29.5
3276	–20.34	–3.10	–0.04	0.94	4.74	1.5660	36.4
3139	–19.29	–2.35	–0.03	0.95	5.01	1.5010	9.3
3013	–18.21	–1.78	–0.03	0.95	5.27	1.4415	35.7
3091	3.59	3.46	–0.02	–0.01	5.73	1.4819	30.1
2951	3.65	3.52	–0.01	–0.01	6.00	1.4147	2.9
2825	3.62	3.48	–0.02	–0.01	6.27	1.3542	29.7

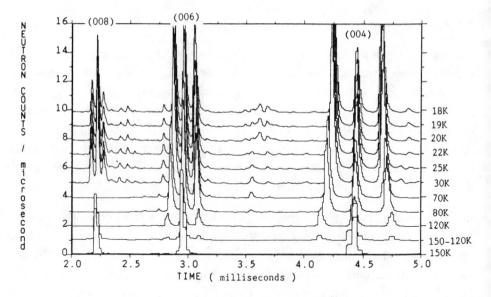

Figure 9 – Temperature variation of the (00l) row in Holmium. Note that the region below 2.5ms was not measured in the scans at 80K and 70K.

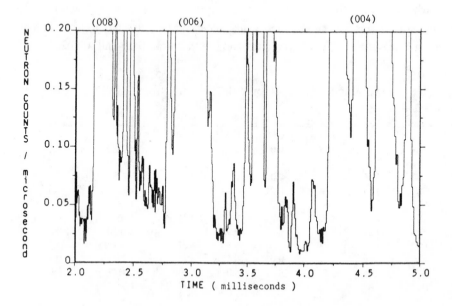

Figure 10 – A section of the (00l) row of Holmium at 18K. The highest peaks on this scale have height 12. This plot shows the richness of the magnetic structure.

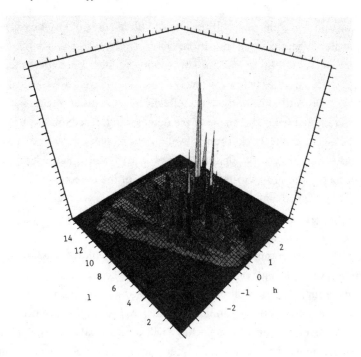

Figure 11 – Reciprocal lattice plots of the (h0l) plane in Holmium (a) above (T = 150K) and (b) below (T = 120K) the magnetic phase transition. The appearance of magnetic satellite peaks is apparent in the low temperature plot.

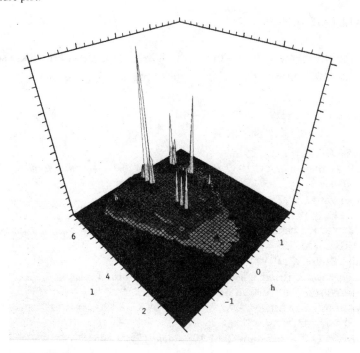

(ii)   Figures 11a and 11b show the (h0l) reciprocal lattice plane of Ho measured above and below the magnetic phase transition. The Q ranges accessed in the two scans were different (different time–of–flight windows), but the appearance of the incommensurate satellites below the phase transition can clearly be seen around the (002), (004), (006) and some of the (10l) reflections. As a point of interest, the locus of the detected region, i.e. the image of the detector in this section through the data set, can be seen in the background level of these plots. Two of these edges correspond to the inner $(1/\lambda_{max})$ and outer $(1/\lambda_{min})$ Ewald sphere surfaces in this crystal geometry, the others to the physical limits of the detector.

6. CONCLUDING REMARKS

The potential of the newly commissioned SXD for performing novel science can only be realised if there is a versatile and comprehensive software package available for use. The data analysis procedures outlined above indicate the data processing route for both standard and non–standard crystallographic problems. As the range of science performed on SXD increases, developments of this package will be essential to keep pace with the data analysis requirements. The foundations of the package discussed above, grounded as they are in the geometry of crystallography, present a solid base from which to build these new developments in the future.

7. ACKNOWLEDGEMENTS

The author wishes to acknowledge all those who have contributed to the SXD software package and those SXD users whose data has been presented in the above.

8. REFERENCES

Anger H O 1958 *Rev. Sci. Insts.* **29** 27
Brown P J and Matthewman J C 1987 *Rutherford Appleton Laboratory Report*
    RAL–87–010
Buras B and Gerward L 1975 *Acta Cryst.* **A31** 372
Busing W R and Levy H A 1967 *Acta Cryst.* **22** 457
Busing W R, Martin K O and Levy H A 1962 ORFLS *Oak Ridge National Laboratory
    Report* ORNL–TM–305
Cooper M J and Rouse K D 1971 *Acta Cryst.* **A27** 622
Forsyth J B, Lawrence R T and Wilson C C 1988 *Nucl. Inst. Meth.* **A273** 741
Forsyth J B and Wilson C C 1990 to be published
Forsyth J B, Wilson C C, Stringer A M, Howard J A K and Johnson O 1986 *J de Phys.
    Colloque C5* **47** 143
Forsyth J B, Wilson C C and Sabine T M 1989 *Acta Cryst.* **A45** 244

Hull S, Goff J, Hutchings M T and Hayes W 1990 to be published
Mair S L and Barnea Z 1971 *Phys. Lett.* **35**A 286
Moze O and Wilson C C to be published
Stirling W G, Tang C C, Haycock P W and Wilson C C 1990 to be published
Wilkinson C 1986 *J de Phys. Colloque C5* **47** 35
Wilkinson C, Khamis H W, Stansfield R F D and McIntyre G J 1988 *J. Appl. Cryst.* **21** 471
Wilson C C 1989 *Rutherford Appleton Laboratory Report* RAL-**89–014**
Wilson C C and Forsyth J B *Rutherford Appleton Laboratory Report* RAL-**89–005**
Wilson C C and Zaleski T 1990 to be published

Inst. Phys. Conf. Ser. No 107: Chapter 3
Paper presented at Neutron Scatt. Data Anal. Conference, Rutherford Appleton, 1990

Reverse Monte Carlo (RMC) simulation: modelling structural disorder in crystals, glasses and liquids from diffraction data

R L McGreevy†, M A Howe†, D A Keen‡ and K N Clausen§

†Clarendon Laboratory, Parks Road, Oxford, OX1 3PU, UK.
‡Neutron Division, Rutherford Appleton Laboratory, Chilton, Didcot, Oxon OX11 0RA
§Risø National Laboratory, Postbox 49, Roskilde, DK-4000 Denmark

Abstract. RMC is a method for producing three dimensional models of the structure of disordered materials that agree quantitatively with the available diffraction data. No interatomic potential is required and data from different sources (neutrons, X-rays, EXAFS) may be combined. The method and its application to neutron diffraction studies of disordered crystals, liquids and glasses will be described.

1. Introduction.

One of the major problems in the analysis of diffraction data on disordered systems has been the lack of any *general* method for producing structural models that agree *quantitatively* with the data. Most analysis is extremely qualitative and based on a few features of the data, e.g. peak positions and coordination numbers derived from radial distribution functions. Monte Carlo and Molecular Dynamics simulations based on an interatomic potential sometimes agree well with experiment (though comparison is normally made with radial distribution functions rather than structure factors), but usually the agreement is only qualitative and occasionally there are major differences. It is not obvious in most cases how the potential should be altered to improve the level of agreement; an iterative procedure is computationally extremely expensive and has only been applied in one or two instances.

The RMC method (McGreevy and Pusztai 1988) overcomes these problems; in this case the structural model is actually fitted to the data and so there must be good agreement (given that the data do not contain significant systematic errors). In this paper we will describe the method in detail and then illustrate some possible applications and ways of analysing or characterising the structures obtained.

2. Metropolis Monte Carlo (MMC).

RMC is a variation of the standard MMC procedure (Metropolis *et al* (1953)). For those unfamiliar with such Monte Carlo methods it is useful to first introduce MMC. The principle is that we wish to produce a statistical ensemble of atoms (configuration) with a Boltzmann distribution of energies. Rather than simply generating and sampling configurations completely at random, which would be a very inefficient procedure, we make use of a weighted sampling procedure (Markov chain) that satisfies certain requirements.

(a) variables $x^{(n)}$ are generated following a rule that requires the $(n+1)^{th}$ element to have a probability distribution $x^{(n+1)}$ that is only dependent on the distribution $x^{(n)}$ of the nth element.

(b) If $P(x \rightarrow y)$ is the probability of reaching state y from state x, then $P(x \rightarrow y)$ must permit movement to every state in the ensemble.

(c) Microreversibility must be satisfied, i.e.

$$x P(x \rightarrow y) = y P(y \rightarrow x)$$

when the system is in equilibrium

For an ensemble in which the number of particles, volume and temperature are fixed (NVT) this may be achieved by the following algorithm.

1. N atoms are placed in a cell with periodic boundary conditions, i.e. the cell is surrounded by images of itself. Normally a cubic cell is used, though other geometries may also be chosen. For a cube of side L the density N/L^3 must equal the required density of the system. The probability of this particular configuration (old = o) is given by

$$P_o \propto e^{-U_o/kT}$$

where U_o is the total potential energy, which may be calculated on the basis of a specified form of the interatomic potential, and T is the specified temperature.

2. One atom is moved at random. The probability of the new (n) configuration is

$$P_n \propto e^{-U_n/kT}$$

and hence

$$\frac{P_n}{P_o} = e^{-(U_n - U_o)/kT} = e^{-\Delta U/kT}$$

3. If $\Delta U < 0$ the new configuration is 'accepted' and becomes the next starting point. If $\Delta U > 0$ then it is accepted with probability P_n/P_o. Otherwise it is 'rejected' and we return to the previous configuration.

4. The procedure is repeated from step 2.

As atoms are moved U will decrease until it reaches an equilibrium value about which it will then oscillate. The maximum size of the random move is normally adjusted so that the ratio of accepted to rejected moves in equilibrium is approximately unity. Configurations are considered to be statistically independent when separated by at least N accepted moves and are then saved. In this way an appropriate ensemble is generated.

3. RMC - the basic method.

In RMC we assume that an experimentally measured structure factor $A^E(Q_i)$ contains only statistical errors that have a normal distribution (this will be discussed in more detail later). The difference between the 'real' structure factor, $A^C(Q_i)$, which can be calculated from a model of the 'real' structure, and that measured experimentally is then

$$e_i = A^E(Q_i) - A^C(Q_i)$$

and has probability

$$p(e_i) = \frac{1}{(2\pi)^{\frac{1}{2}}\sigma(Q_i)}\exp(-e_i^2/2\sigma^2(Q_i)))$$

where $\sigma(Q_i)$ is the standard deviation of the normal distribution. The total probability of A^C is

$$P = \prod_{i=1}^{m_Q} p(e_i) = \left(\frac{1}{(2\pi)^{\frac{1}{2}}\overline{\sigma}}\right)^{m_Q} \exp\left[-\sum_{i=1}^{m_Q} e_i^2/2\sigma^2(Q_i)\right]$$

where m_Q is the number of Q_i points in A^E and

$$\overline{\sigma} = \left(\prod_{i=1}^{m_Q} \sigma(Q_i)\right)^{1/m_Q}.$$

In order to model the structure of a system using A^E we therefore wish to create a statistical ensemble of atoms whose structure factor satisfies the above probability distribution. Writing the exponent as

$$\chi^2 = \sum_{i=1}^{m_Q} [A^C(Q_i) - A^E(Q_i)]^2/\sigma^2(Q_i)$$

then $P \propto \exp(-\chi^2/2)$ and it can immediately be seen that $\chi^2/2$ in RMC is equivalent to U/kT in MMC. The algorithm for RMC is therefore as follows.

1. Start with an initial configuration with periodic boundary conditions. Calculate

$$g_o^C(r) = \frac{n_o^C(r)}{4\pi r^2 dr\rho}$$

where ρ is the atom number density and $n^C(r)$ is the number of atoms at a distance between r and $r+dr$ from a central atom, averaged over all atoms as centres. Transform to

$$A_o^C(Q) - 1 = \rho \int_0^\infty 4\pi r^2 (g_o^C(r) - 1)\frac{\sin Qr}{Qr}dr.$$

Calculate

$$\chi_o^2 = \sum_{i=1}^{m} [A_o^C(Q_i) - A^E(Q_i)]^2/\sigma^2(Q_i)$$

2. Move one atom at random. Calculate the new $g_n^C(r)$ and $A_n^C(Q)$ and

$$\chi_n^2 = \sum_{i=1}^{m} [A_n^C(Q_i) - A^E(Q_i)]^2/\sigma^2(Q_i)$$

3. If $\chi_n^2 < \chi_o^2$ the move is accepted. If $\chi_n^2 > \chi_o^2$ the move is accepted with probability $\exp(-(\chi_n^2 - \chi_o^2)/2)$. Otherwise it is rejected.

4. Repeat from step 2.

χ^2 will initially decrease until it reaches an equilibrium value about which it will oscillate. The resulting configuration should be a three dimensional structure that is consistent with the experimental structure factor within the experimental error. Independent configurations may then be collected under the same conditions as for MMC.

where $\sigma(Q_i)$ is the standard deviation of the normal distribution. The total probability of A^C is

$$P = \prod_{i=1}^{m_Q} p(e_i) = \left(\frac{1}{(2\pi)^{\frac{1}{2}}\overline{\sigma}}\right)^{m_Q} \exp\left[-\sum_{i=1}^{m_Q} e_i^2/2\sigma^2(Q_i)\right]$$

where m_Q is the number of Q_i points in A^E and

$$\overline{\sigma} = \left(\prod_{i=1}^{m_Q} \sigma(Q_i)\right)^{1/m_Q}.$$

In order to model the structure of a system using A^E we therefore wish to create a statistical ensemble of atoms whose structure factor satisfies the above probability distribution. Writing the exponent as

$$\chi^2 = \sum_{i=1}^{m_Q} [A^C(Q_i) - A^E(Q_i)]^2/\sigma^2(Q_i)$$

then $P \propto \exp(-\chi^2/2)$ and it can immediately be seen that $\chi^2/2$ in RMC is equivalent to U/kT in MMC. The algorithm for RMC is therefore as follows.

1. Start with an initial configuration with periodic boundary conditions. Calculate

$$g_o^C(r) = \frac{n_o^C(r)}{4\pi r^2 \, dr \rho}$$

where ρ is the atom number density and $n^C(r)$ is the number of atoms at a distance between r and $r+dr$ from a central atom, averaged over all atoms as centres. Transform to

$$A_o^C(Q) - 1 = \rho \int_0^\infty 4\pi r^2 (g_o^C(r) - 1)\frac{\sin Qr}{Qr} dr.$$

Calculate

$$\chi_o^2 = \sum_{i=1}^{m} [A_o^C(Q_i) - A^E(Q_i)]^2/\sigma^2(Q_i)$$

2. Move one atom at random. Calculate the new $g_n^C(r)$ and $A_n^C(Q)$ and

$$\chi_n^2 = \sum_{i=1}^{m} [A_n^C(Q_i) - A^E(Q_i)]^2/\sigma^2(Q_i)$$

3. If $\chi_n^2 < \chi_o^2$ the move is accepted. If $\chi_n^2 > \chi_o^2$ the move is accepted with probability $\exp(-(\chi_n^2 - \chi_o^2)/2)$. Otherwise it is rejected.

4. Repeat from step 2.

χ^2 will initially decrease until it reaches an equilibrium value about which it will oscillate. The resulting configuration should be a three dimensional structure that is consistent with the experimental structure factor within the experimental error. Independent configurations may then be collected under the same conditions as for MMC.

The algorithm used here is not strictly statistically correct, since we are actually sampling χ^2 (by varying A^C) and not the data (by multiple measurements of A^E). We should therefore use

$$\left(\frac{\chi_n^2}{\chi_o^2}\right)^{m_Q/2-1} \exp(-(\chi_n^2 - \chi_o^2)/2)$$

in place of $\exp(-(\chi_n^2 - \chi_o^2)/2)$. However, given that most experimental data do not contain only statistical errors (see section 5.5) and the loss of direct analogy with MMC, the former algorithm has been preferred at present. Howe (1989) has successfully used the far simpler criterion of accepting 5% of all moves that increase χ^2.

4. RMC - variations on the basic method.

RMC is easily adapted so that different functions are used. The basic method outlined above is for modelling the structure factor of a one component system. Instead of modelling $A^E(Q_i)$ we could alternatively model $g^E(r_j)$ (for reasons to be discussed later the two are *not* identical), and χ^2 becomes

$$\chi^2 = \sum_{i=1}^{m_r} [g^C(r_j) - g^E(r_j)]/\sigma^2(r_j)$$

This is obviously computationally faster since a transform is not required during each iteration. For a multicomponent system where different total structure factors (indicated by index k)

$$F_k^E(Q_i) = \sum_{\alpha,\beta} c_\alpha c_\beta b_\alpha b_\beta (A_{\alpha\beta}(Q_i) - 1)$$

have been measured, for instance using neutron diffraction with isotopic substitution, we then have

$$\chi^2 = \sum_k \sum_{i=1}^{m} [F_k^C(Q_i) - F_k^E(Q_i)]/\sigma_k^2(Q_i)$$

c_α is the concentration and b_α the coherent scattering length of component α; $A_{\alpha\beta}(Q)$ are the partial structure factors. In the same way results from neutron and X-ray diffraction measurements may be combined (McGreevy 1989). In fact any data set may be used if a spectrum S^C can be calculated from the structure and compared to the experimental S^E. This has been done for EXAFS (Gurman and McGreevy (1990)) and may be possible in some cases for NMR.

5. RMC - simulation details.

5.1. Configuration size.

When starting from the initial configuration $g(r)$ must be calculated. This involves a summation of order N^2. However for each particle move it is only necessary to calculate the *change* in $g(r)$ corresponding to the moved particle, which is a summation of order N. This is the same in MMC, but not in Molecular Dynamics (MD) where all moves are of order N^2 (unless the potential is truncated). For this reason MC simulations may involve much larger configurations than are used in MD. Generally we use $N > 1500$, and have used $N \approx 30,000$. The size of simulation is important when modelling $A^E(Q_i)$ since $g^C(r_j)$ may only be calculated up to $r = L/2$. In order to be able to transform $g^C(r_j)$ directly to $A^C(Q_i)$ we require that $g^C(r > L/2) = 1$. Any significant deviations from this, either due to long range correlations or statistical fluctuations, will cause truncation ripples at low Q in $A^C(Q_i)$. Size is also relevant when modelling $g^E(r_j)$, because this determines the statistical fluctuations in $g^C(r_j)$ and hence the effective value of $\sigma(r_j)$.

5.2. r spacing.

The minimum r spacing, when modelling $A^E(Q)$, is determined by the Q range, i.e. the value of Q_{m_Q}. The real space resolution is then $2\pi/Q_{m_Q}$; one requires ≈ 5 points over this range so an r spacing of $2\pi/5Q_{m_Q}$ is appropriate. For simple liquids where structure in $A^E(Q)$ extends out to $Q \approx 10\text{Å}^{-1}$ this makes a spacing of $\approx 0.1\text{Å}$ suitable, whereas for molecular liquids or glasses with structure out to $\approx 40\text{Å}^{-1}$ a spacing of $\approx 0.025\text{Å}$ is suitable. However it should be noted that decreasing the r spacing for a fixed number of particles increases the statistical error in $g^C(r_j)$, so it may be necessary to increase the model size. Another alternative might be to use a non-uniform r spacing, with larger r spacing at large r.

5.3. Closest approach distance of two atoms.

If one has perfect data then the distances of closest approach of two atoms are determined by the low r cut-off in $g^E(r_j)$. However for imperfect data, particularly when $A^E(Q_i)$ is significantly truncated at the maximum Q value, the closest approach may not be well defined. For this reason it is usually sensible to specify allowed distances of closest approach, i.e. to define an excluded volume. This also saves considerable time since moves which would result in atoms being too close together can be rejected before calculation of the change in $g^C(r_j)$. For good data the specified closest approaches may be somewhat lower than the 'real' values but for poor data they need to be more carefully chosen. However if the values are too large then this is usually apparent because the resulting $g^C(r_j)$ has a sharp cut-off instead of decreasing more gradually to zero. If they are too low $g^C(r_j)$ may have a sharp 'spike' in the low r region.

While this is a very simple constraint on the structure it is also very powerful, since the imposition of both an excluded volume and a fixed density restricts possible configurations. One could also view it as the imposition of a hard sphere repulsive potential. In the case of a two component system where the hard sphere radii are sufficiently different one can attempt to obtain all three partial radial distribution functions or structure factors from one or two total structure factors, while three are required for a conventional solution. This is valuable in cases where suitable isotopes are not available for neutron diffraction. Since the resulting structure is then dependent on the choice of radii one should, if possible, make a choice based on other experimental information rather than treating them as free parameters.

5.4. Maximum size of random move.

The maximum size, d, of the random move determines the ratio of accepted to rejected moves, but also determines the amount that the structure may change with each move. If d is too small then nearly all moves will be accepted but the structure will change little, while if it is too large then few moves will be accepted and the average structural change will also be small. If we attempt to choose d such that the ratio of accepted to rejected moves is ≈ 1, as in MMC, this often leads to a value $d < 0.1\,\text{Å}$. The average structural change per move is usually maximised for $0.1 < d < 0.3\,\text{Å}$ with an acceptance/rejection ratio ≈ 0.5, so this range is normally used.

When starting from a structure that is significantly different from the 'real' structure it is possible that certain atoms become 'trapped' in some local arrangement. This is a local minimum in the minimisation procedure, rather than the required global minimum. One way around this problem is to run the simulation for a while with a large value of d, for instance up to 10 Å. While hardly any moves will be accepted those that are may be sufficient to get the configuration out of the local minimum. (The terms 'local' and 'global' here are used here in an unconventional manner; 'global' refers to *any* minimum that satisfies the required fitting criterion and 'local' to any minimum that does not. We specifically *do not* wish to attain the conventional global minimum, i.e. the single structure that is closest to the data.)

5.5. Experimental error -σ

The RMC algorithm assumes that we have only statistical errors. In practice this is not true, but the whole procedure is not thereby invalidated. A three dimensional structure that is consistent with the experimental data within *some measure* of the error can still be produced, though this measure is now less well defined.

A real experimental structure factor $A^E(Q_i)$ will contain both statistical and systematic errors. While one might expect statistical errors to be small where $A^E(Q_i)$ is large, and vice-versa, in practice the requirement to perform container and background corrections in many experiments means that statistical errors are often quite uniformly distributed. In many X-ray experiments counting times are chosen to deliberately produce a uniform distribution. Since we often have no knowledge of the likely distribution of systematic errors it is usually simplest to assume a constant value of σ at all Q_i, though σ may differ between different data sets. However there have been cases in which large values of $\sigma(Q_i)$ have been used in particular Q ranges where it is known that there were errors in the data. By setting $\sigma(Q_i)$ at an extremely large value these errors can effectively be ignored.

When $A^E(Q_i)$ is transformed to $g^E(r_j)$ the errors are 'redistributed'. There are also statistical errors inherent in $g^C(r_j)$. Comparison at the $g(r)$ stage will therefore not produce precisely the same result as comparison at the $A(Q)$ stage. This is of course more exaggerated when additional errors are introduced in $g^E(r_j)$ by truncation of $A^E(Q_i)$. It is also worth noting that certain features of $A^E(Q_i)$, in particular 'pre-peaks' or 'first sharp diffraction peaks' in some glasses and liquids at $Q \approx 1\,\text{Å}^{-1}$, correspond to small, long period modulations of $g^E(r_j)$. It is possible that such real space structure can be 'ignored' to a great extent if the modulation amplitude is comparable to the chosen value of $\sigma(r_j)$. For these reasons it is strongly recommended that RMC is used for modelling $A^E(Q_i)$ wherever possible.

From the above discussion it is clear that the precise value of σ is not known in any particular case; it may therefore be considered as a parameter of the simulation. If we make the analogy with MMC that $\chi^2 = U/kT$ then $\sigma = kT$. Under normal circumstances we would use $\sigma \approx 1\%$, i.e. a typical value of experimental error. However if it is believed that the simulation has run into a local minimum then, as is common procedure in other simulations, we would increase the value of σ, i.e. increase the temperature. After running the simulation for a while σ would then be decreased to its original value. Alternatively if we *deliberately* wished to find a local minimum *closest* to a particular starting configuration then we can effectively set $\sigma = 0$ by only accepting moves which decrease χ^2.

We have generally found for disordered structures that the global minimum in χ^2 is relatively broad and little manipulation of σ or d is required to reach it. However for highly ordered structures (such as crystals) this would not be the case and global minimisation of χ^2 would only be achieved by simulated annealing with σ as the control parameter.

5.6. Renormalisation.

Experimental data will normally contain small normalisation errors in the form of multiplicative and additive constants. It is possible to take account of such errors within the RMC algorithm (Howe (1989)). This can be particularly important when dealing with isotopic substitution neutron diffraction data when the relative normalisation of different structure factors must be correct. The required multiplicative factor α which minimises χ^2 is given by

$$\alpha = \frac{\sum_{i=1}^{m_Q} A^E(Q_i)A^C(Q_i)}{\sum_{i=1}^{m_Q} (A^E(Q_i))^2}$$

It is recommended that such renormalisation only be performed as a refinement when (a) α is close to 1 and (b) χ^2 is small, i.e. a reasonable fit has already been achieved. If the required value of α differs significantly from 1 then the experimental data should obviously be checked anyway. If χ^2 is large then one effect of renormalisation may be to simply decrease χ^2 by making α small and $A^C(Q_i)$ a uniform distribution.

The additive factor β which minimises χ^2 is given by

$$\beta = \frac{1}{m_Q} \sum_{i=1}^{m_Q} (A^E(Q_i) - A^C(Q_i))$$

It is generally safe to use this factor for all data though again if the value is large the original data should be checked.

When α and β factors are used simultaneously the formulae are more complex - see Howe (1989) for details.

6. RMC - why use it?

There are numerous methods of structural modelling, from the simplest 'hand built' or 'hand drawn' models to conventional MC or MD simulations. However RMC has several advantages.

1. RMC uses all the available structural data, not just particular features, in a quantitative rather than qualitative manner. Many models that use particular features, e.g. peak positions and coordination numbers from radial distribution functions, can be misleading.

2. RMC is potential independent. If a potential exists that, when used in an MC or MD simulation, also produces quantitative agreement with experimental structure factors, then this is obviously equally as good as RMC (and one would hope that the results were similar!). However few potentials provide such quantitative agreement, and in some cases it has not yet proved possible to produce potentials that provide qualitative agreement. Also most simulations are compared to experiment at $g(r)$ stage because the configurations are too small to allow transformation to $A(Q)$. For the reasons discussed above it is important for good structural modelling that comparison be made at the $A(Q)$ stage.

3. Because RMC models a three dimensional structure $g^C(r_j)$ and $A^C(Q_i)$ must correspond to a possible physical structure. However $g^E(r_j)$ derived by conventional methods may contain errors which mean that it *could not* correspond to a possible physical structure; i.e. it is internally inconsistent. In the case of multicomponent systems there is no requirement that the partial radial distribution functions derived by conventional methods are consistent with one another, while in RMC they must be consistent and physically possible. This constraint 'improves' the separation of partials in cases where the separation matrix is poorly conditioned.

4. Different types of data, e.g. neutron and X-ray diffraction, can be modelled. The different data sets can have different Q ranges, spacings, resolutions etc. It is also easy to include additional constraints on the structure; these could be from other experimental information (e.g. NMR) or could be some other 'knowledge' of the system (e.g. chemical bonding ideas).

5. RMC is easily adapted to different physical problems.

7. Uniqueness.

The three dimensional structure produced by RMC is *not* unique, it is simply a model that is consistent with the data and any additional constraints. Other methods that produce structures which are equally consistent with the data are equally valid and there is no way of determining which is 'correct' *in the absence of any additional information*. One possible disadvantage of RMC is that it tends to produce the most disordered structure that is consistent with the data and constraints, i.e. the configurational entropy is maximised. However this is counteracted by the ability to include additional constraints, which means that additional ordering can be imposed and a range of consistent structures investigated; those that are found to be inconsistent can then be discounted. We are at present investigating methods of including constraints that are non-specific, i.e. they will decrease the configurational entropy without specifying how it should be decreased.

In the special case of a system for which the interatomic potential is *purely* pairwise additive there is a theoretical justification for the determination of the three dimensional structure from a one dimensional $g(r)$ or $A(Q)$ (Evans (1989)). Given that the potential uniquely determines the structure

$$\phi(r) \rightarrow g^{(2)}(r_1, r_2), g^{(3)}(r_1, r_2, r_3), g^{(4)}(...)$$

where $g^{(2)}(r_1, r_2) = g(r)$ and $g^{(n)}(...)$ are the n-body correlation functions, then for a pairwise additive potential there is a *functional* relationship between $\phi(r)$ and $g(r)$ such that

$$g^{(2)}(r_1, r_2) \rightarrow \phi(r)$$

i.e. $g^{(2)}$ uniquely determines ϕ. (This is not to say that we can write down the relationship, but merely that one exists.) If $g(r)$ determines ϕ and ϕ determines the structure, then $g(r)$ determines the structure i.e.

$$g(r) \rightarrow g^{(3)}(r_1, r_2, r_3), g^{(4)}(...)$$

While the potentials in real systems are never purely pairwise additive (though such potentials are used in the majority of MC and MD simulations) the above result does indicate that a precisely measured $g(r)$ or $A(Q)$ does contain a *great deal* of information about the three dimensional structure. RMC is one possible way of attempting to extract this information.

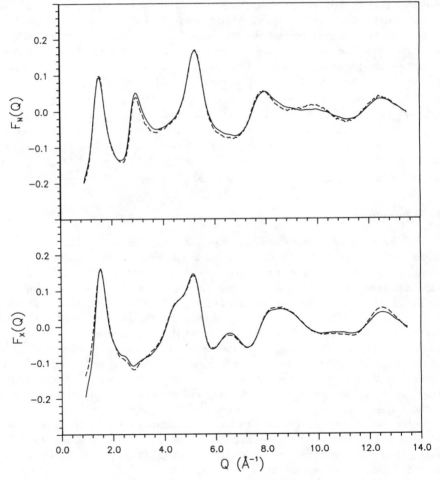

Figure 1. Structure factors for vitreous silica: neutron diffraction (upper graph) and X-ray diffraction (lower graph). Broken line - experiment; solid line - RMC fit.

8. Applications of RMC.

8.1. *The structure of vitreous silica*

Neutron and X-ray diffraction patterns of vitreous silica have been modelled by Keen and McGreevy (1990) using RMC. The experimental and fitted structure factors are shown in figure 1. This study illustrates the ability of RMC to combine different

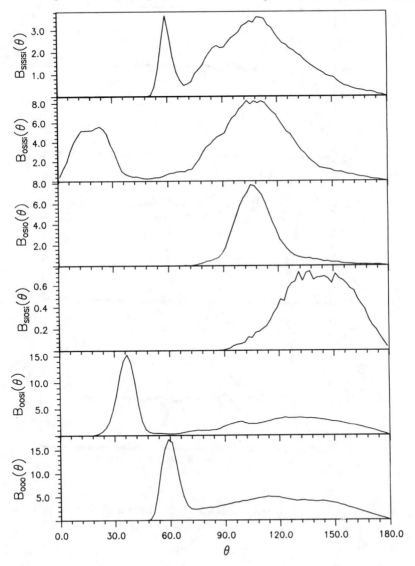

Figure 2. Bond angle distributions for vitreous silica. Maximum bond lengths are defined by the first minima in the appropriate partial radial distribution functions.

types of experimental data and to determine three partial radial distribution functions and three partial structure factors for a two component system from two independent measurements, rather than the three required for a conventional analysis. (This is not to say that three measurements would not be preferable since any additional data or constraints limits the range of consistent structures.) In addition RMC produces a three dimensional structure which may be analysed further, for example by calculation of bond angle distributions (figure 2).

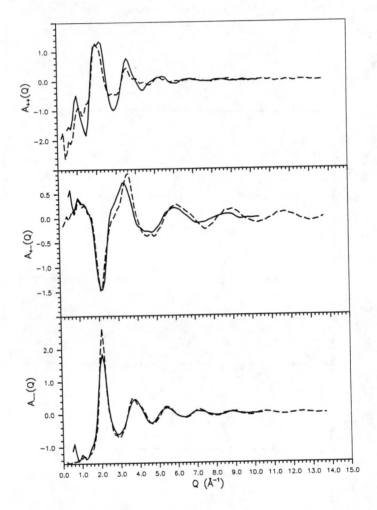

Figure 3. Partial structure factors for molten (solid line) and vitreous (broken line) $ZnCl_2$.

8.2. *Intermediate range order in* $ZnCl_2$ *melt and glass*

In this case the three partial radial distribution functions for molten $ZnCl_2$ (obtained using neutron diffraction with isotopic substitution) have been fitted (McGreevy and Pusztai (1990)) and the resulting structure then used as the starting point for modelling the total structure factor of $ZnCl_2$ glass (no isotopic substitution) (Pusztai and McGreevy (1990)). Since the structures of glass and liquid are expected to be similar this maximises the information that can be obtained from the glass data and three partial structure factors are obtained from a single measurement.

Both glass and liquid have a peak in the partial structure factor $A_{ZnZn}(Q)$ at $Q \approx$ 1 $Å^{-1}$ (figure 3) which is characteristic of intermediate range ordering, though the nature of this order has not been understood. Interestingly the peak is significantly smaller in the glass than in the liquid. Figure 4 shows a section from an RMC generated

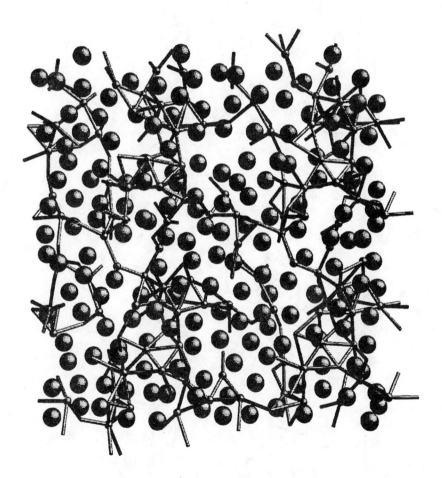

Figure 4. Section from a configuration for molten $ZnCl_2$ showing Zn^{2+} ions (small) and Cl^- ions (large) and Zn-Zn bonds. Bonds are drawn from ions in the section to all others, so some bonds terminate on ions outside the section.

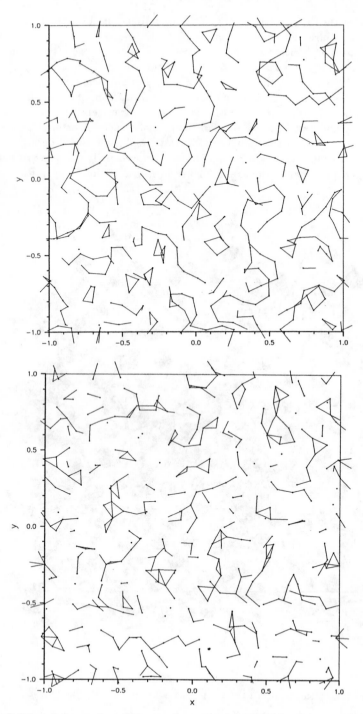

Figure 5. S-S bonds in sections from configurations for liquid sulphur at 423K: high 2-fold coordination (upper graph) and low 2-fold coordination (lower graph).

configuration for liquid $ZnCl_2$. Zn-Zn bonds are drawn where a bond is defined as a vector joining two Zn ions whose length is less than the position of the first minimum in $g_{ZnZn}(r)$. It is clear that there are 'clusters' of bonds connected by 'chains', giving a non-uniform bond density. It is this fluctuation in the Zn^{2+} ion distribution that gives rise to the 1 Å^{-1} peak in $A_{ZnZn}(Q)$.

8.3. Rings and chains in liquid sulphur

There is considerable controversy at present concerning the structure of liquid sulphur; is it based on rings, chains or a mixture and what changes in structure occur at the viscous transition? RMC fitting of the structure factor at 423K and 503K, either side of the transition, gives a structure of 'entangled broken chains' (using the description of liquid tellurium given by Hafner (1990)). Though the average coordination of S atoms is close to 2 only $\approx 40\%$ are actually 2 coordinated with $\approx 40\%$ being 1 or 3 coordinated. By constraining RMC it is possible to produce a structure that is 92% 2 coordinated and yet equally consistent with the data. Sections from configurations with low and high 2 coordination are shown in figure 5. The latter has considerably longer chains. This difference is most clearly illustrated by a self-avoiding random walk analysis of the structure; the distribution of path lengths is shown in figure 6. Interestingly there are slightly more very long chains at 503K than 423K as would be expected for a higher viscosity.

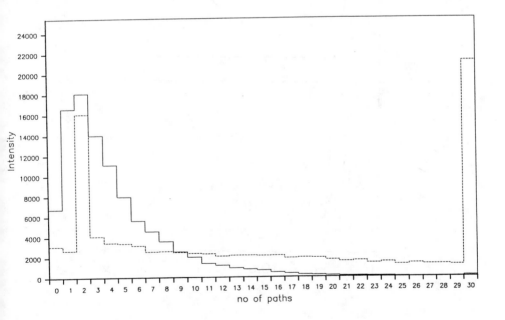

Figure 6. Self-avoiding random walk analysis for the configurations illustrated in figure 5. Broken line - high 2-fold coordination, solid line - low 2-fold coordination.

The complete absence of rings from these configurations does not imply that rings may not exist, since ring structures have a lower configurational entropy than chains and hence are not favoured by unconstrained RMC. It is known that S_6 and S_8 ring molecules on their own do not agree with the structure factor but more complex rings may do; this is currently being investigated.

8.4. Disorder at the approach to melting in AgBr

This has been studied by RMC analysis of powder neutron diffraction data (Keen, McGreevy and Hayes (1990); Keen, Hayes, McGreevy and Clausen (1990)). Rather than a conventional crystallographic analysis of Bragg peaks RMC is used to analyse the total structure factor, i.e. Bragg and diffuse scattering simultaneously. The model

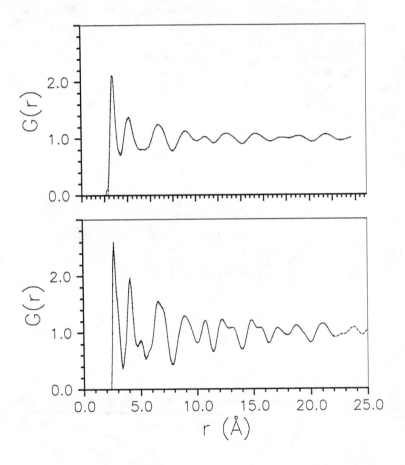

Figure 7. Total radial distribution functions $G(r)$ for AgBr at T=490 (lower graph) and T=699 K (upper graph). Broken line - experiment; solid line - RMC fit.

thus reflects both the long range order and short range disorder that produce the diffraction pattern. Because the correlation lengths inherent in Bragg peaks are very large it is not possible to fit the structure factor $F(Q)$ directly - it must be transformed to give a radial distribution function $G(r)$. We therefore require that $F(Q)$ is measured over a sufficiently wide Q range to avoid truncation errors at both low and high Q. Crystallinity is allowed for within the periodic boundary conditions if $L = na$ where n is an integer and a is the lattice constant.

RMC fits to $G(r)$ for AgBr at two temperatures are shown in figure 7. In this particular model $n = 8$, i.e. the simulation configuration contains 8^3 unit cells. The density distribution in the (cubic) crystallographic unit cell can then be calculated by averaging over all these individual unit cells (and also over multiple configurations). A cut through this distribution in the $< 111 >$ direction across a Ag^+ lattice site is shown in figure 8. As T increases towards the melting point a peak in density grows at the $(\frac{1}{4}, \frac{1}{4}, \frac{1}{4})$ interstitial site; integration across this peak can be used to obtain the number of interstitials. Analysis of the density distribution along different paths can also be used to give information on ion conduction mechanisms.

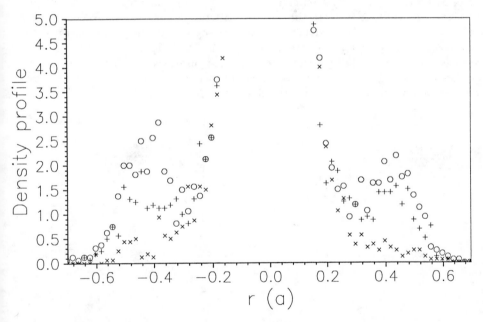

Figure 8. The density profile across Ag^+ lattice sites in $< 111 >$ directions at T = 669 ($\times$), 689 ($+$) and 699K ($\circ$).

8.5. Defects in yttria doped zirconia

When zirconia (ZrO_2) is doped with yttria (Y_2O_3), anion vacancies are introduced into the material and defect clusters form as the crystal structure relaxes around them. These relaxations correspond to regions in reciprocal space where diffuse scattering is observed. This diffuse scattering has been measured with neutrons for a number of Y_2O_3 dopant concentrations using single crystal samples. Most analyses of this type of scattering involve the adoption of a specific defect cluster from which the expected diffuse scattering is calculated and if agreement with the data is poor, the cluster is adapted until the agreement is improved. Hence the defect cluster, although it may involve many ions, is of only one type. Problems also occur when the defect concentration increases and these clusters start to interact. We have used a form of the RMC technique to introduce more flexibility into the existing models. Starting from a configuration containing the previous 'best-fit' defect clusters, we have randomly switched between ion types (regular site, vacancies or defects) on random lattice points (i.e. a 'lattice gas' model) and compared the resulting calculated 'diffuse' structure factor with the data directly after each move until χ^2 is minimised. Using this method speeds up the calculation since the 'diffuse' structure factor may be calculated as a difference between the total structure factor (Bragg and diffuse scattering) and the ideal structure factor (Bragg scattering only). The resulting fits are still at an early stage and we have not analysed the resulting configurations but the agreement with the data is encouraging (figure 9). A complete description of the method and analysis of the structure produced by this method will be published later.

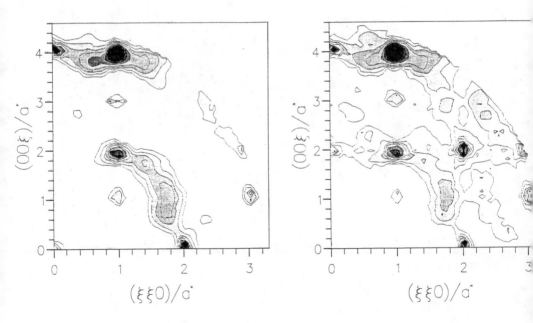

Figure 9. Diffuse scattering for 9 mol% Y_2O_3 doped ZrO_2: experiment (left) and RMC lattice gas model fit (right).

8.6. Other applications

Other applications of RMC to date are

Condensed inert gases (McGreevy and Pusztai (1988))
Liquid metals (unpublished)
Liquid alkali-lead alloys (Howe and McGreevy (1990))
Molten salts (McGreevy and Pusztai (1990))
Liquid CuSe (Howe (1989))
Liquid halogens (Howe (1990a))
$NiCl_2$ solution (Howe (1990b))

9. Conclusions

RMC is a powerful method for producing structural models of disordered systems based on diffraction data. Its main disadvantage as a data analysis technique is the large amount of CPU time required and for this reason many experimentalists may feel that it could not be applied to their own data. However during the workshop we have modelled the structure factor for liquid bismuth (figure 10) using a 80386 based PC

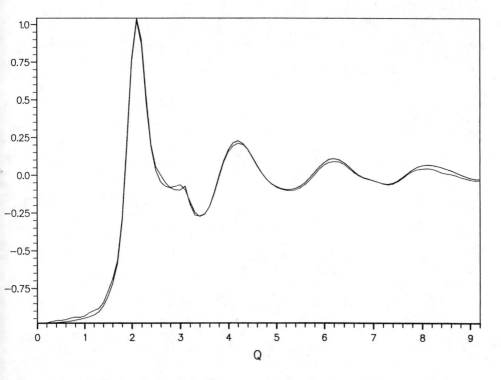

Figure 10. Structure factor for molten Bi. Broken line - experiment; solid line - RMC fit.

with a single T800 transputer, in less than 12 hours (the speed of this system is actually 0.6 of a VAX 8700). The *total* system cost was less than £4000 and the program was written in standard Fortran. This type of analysis could therefore be used by nearly all experimental groups.

RMC programs for some applications are available from the authors (Clarendon Laboratory).

References

Evans R 1989 Mol. Simulation

Gurman S J and McGreevy R L 1990 In preparation

Hafner J 1990 J. Phys: Condensed Matter **2** 1271

Howe M A 1989 Physica B **160** 170

Howe M A 1990a Mol. Phys. **69** 161

Howe M A 1990b J. Phys: Condensed Matter **2** 741

Howe M A and McGreevy R L 1990 In preparation

Keen D A and McGreevy R L 1990 Nature **344** 423

Keen D A, McGreevy R L and Hayes W 1990 J. Phys: Condensed Matter **2** 2773

Keen D A, Hayes W, McGreevy R L and Clausen K N Phil. Mag. Lett. in press

McGreevy R L 1989 IOP conference series **101** 41

McGreevy R L 1990 Nuovo Cimento in press

McGreevy R L and Pusztai L 1988 Mol. Simulation **1** 359

McGreevy R L and Pusztai L 1990 submitted to Proc. Roy. Soc. A

Metropolis N, Rosenbluth A W, Rosenbluth M N, Teller A H and Teller E 1953 J. Phys. Chem. **21** 1087

Pusztai L and McGreevy R L 1990 J. Non-cryst. Solids in press

Inst. Phys. Conf. Ser. No 107: Chapter 3
Paper presented at Neutron Scatt. Data Anal. Conference, Rutherford Appleton, 1990

185

A normalization procedure of liquid diffraction data based on the knowledge of the sample neutron cross section

F. Cilloco and R. Felici

Istituto di Struttura della Materia, Via E. Fermi 38, I-00044 Frascati, Italy

ABSTRACT: We report on a method to normalize neutron scattering data (accuracy of the order of 1%) from liquids based on an accurate knowledge of the total scattering cross section. This procedure can be used when it is not possible or it is unlikely to perform a normalization run. The method has been developed for continuous sources but it can be in principle applied to the pulsed neutron sources case. Some experimental results, obtained using the presented normalization procedure, will be presented and discussed.

1. INTRODUCTION

The scope of a diffraction measurement from a liquid sample is the knowledge of the structure function S(Q) [1], Q being the exchanged momentum vector,which contains all the information regarding the structure of the sample. Unfortunately the S(Q) is not directly accessible from diffraction data, but many different corrections must be applied in order to extract this function correctly.

Among others factors, the procedure to normalize the intensity data to the correct differential cross section assumes particular importance [2]. In general this operation can be performed by comparing the spectra obtained from the sample with an independent incoherent spectra under the same experimental conditions. For this purpose a sample of polycrystalline vanadium, which is an almost perfect incoherent scatterer, is used. Unfortunately this means that we must apply the whole range of data reduction on both scattering data sets and put particular attention on conserving the same experimental conditions (sample volume, shields if any etc.).

There are, however, occasion in which we need only a rough quick estimation of differential cross section, or to have an independent check in doubtful cases or situations when it is impossible or very hard to perform a correct vanadium experiment, or simply because we need a simpler analytical correction.

In this paper we report on a first effort in this direction from a theoretical point of view and we present some results which seem very encouraging. Our procedure still needs extensive applications to different class of systems to learn more about its general applicability, as well as a possible extension to treat data from TOF machines

2. THEORY

The proposed method is based on the knowledge of the total scattering cross section, which is easily obtainable by a transmission measurement for neutrons of the same energy of those used in the main experiment.

The cross section contains in integral form all the information regarding the scattering process. In general, for a steady state neutron source, what we measure is the number of neutrons arriving at the detector at different fixed angles in defined time units. Considering the case of a black detector and after having corrected for background and for the sample self absorption, the intensity can be written as:

$$1) \qquad I(\vartheta) = A(\vartheta)\, k(\vartheta)\, \rho \left[\frac{d\sigma_{coh}}{d\Omega} + \frac{d\sigma_{inc}}{d\Omega} + P(\vartheta) + \frac{\sigma_E}{4\pi} m(\vartheta) \right]$$

where:

ϑ is the scattering angle,

$A(\vartheta)$ is the self-absorption factor,

$k(\vartheta)$ is the normalization constant, in the following supposed to be angle independent

$P(\vartheta)$ is the term which include all inelastic effects,

σ_E is the experimental total cross section,

$m(\vartheta)$ is the percentage of multiple scattering events. For simplicity we outline the normalization procedure for constant multiple scattering, simply substituting $m(\vartheta)$ with m.

The difference between scattering cross section which is measured with neutrons of a given energy with respect to the bond scattering cross section is due to the contribution of the inelastic scattering. We may so write:

$$2) \qquad \Delta\sigma = \sigma_B - \sigma_E \approx 2\pi \int_0^\pi k\, \rho\, P(\vartheta) \sin(\vartheta)\, d\vartheta$$

in which σ_B is the bound scattering cross section readily available in tabulated form. On the other hand we must have, ϑ being the scattering angle:

3) $\qquad 2\pi \int_0^\pi I(\vartheta) \sin(\vartheta) d\vartheta = k \rho \sigma_E(1 + m)$

If we can get measurements over the whole scattering angle (i.e. up to $Q=4\pi/\lambda$) from equation (3) the instrumental constant k is readily obtainable after an integration over ϑ.

However, in general, only a limited range of ϑ is experimentally accessible ($\vartheta_{MIN} \leq \vartheta \leq \vartheta_{MAX}$) so we must compensate for the lack of knowledge from 0 to Q_{MIN} ($Q_{MIN} = 4\pi/\lambda \sin(\vartheta_{MIN}/2)$) and from Q_{MAX} to $4\pi/\lambda$ ($Q_{MAX} = 4\pi/\lambda \sin(\vartheta_{MAX}/2)$). For the small angle part of the measured spectra an interpolation to $\vartheta=0$ is in general easy and quite sufficient as approximation. In fact we note that because of the presence of the sin function in equation (3), the weight of the low Q region of the curve is less significant than the central part, helping in the choice of a simple approximation.

For the high Q part we instead proceed as follow.

We use an approximate form of the inelastic correction, given by the first order Placzeck expansion [3,4,5]. The first two terms can be written as:

4) $\qquad P(Q) = A + B Q^2$

with:

4b) $\qquad A = \dfrac{m}{2M} \dfrac{k_B T}{E_{inc}} \qquad\qquad\qquad B = \dfrac{m}{2M} f\left(\dfrac{k_B T}{E_{inc}}\right)$

Neglecting the constant A which is generally of the order of some tenth of percent, we can use the expression (2) to obtain the coefficient B without any knowledge of the dynamic structure factor $S(Q,\omega)$ and of its moments. We note that the inelastic component can be obtained introducing equation (4) in equation (2):

5) $\qquad \Delta\sigma = 2\pi \left(\dfrac{4\pi}{\lambda}\right)^2 B \int_0^\pi \sin\left(\dfrac{\vartheta}{2}\right) \sin(\vartheta) d\vartheta$

from which:

6) $$B = \frac{\lambda^2}{32\,\pi^3}\,\Delta\sigma$$

We have assumed that for $Q_{MAX} \le Q \le 4\pi/\lambda$ the coherent oscillations of the $S(Q)$ can be neglected. As a consequence of this assumption the scattered intensity in this range can be described by the intensity of an incoherent scatterer with the same scattering cross sections (σ_B and σ_E) and same inelastic correction of our sample and we may write:

7) $$I_P(\vartheta) = k\,\rho \left[\frac{\Delta\sigma}{2\pi}\left(\sin\left(\frac{\vartheta}{2}\right)\right)^2 + \frac{\sigma_B}{4\pi} + m(\vartheta)\frac{\sigma_E}{4\pi} \right]$$

so that equation (3) can be split and becomes:

8) $$\int_0^{\vartheta_{MAX}} I(\vartheta)\sin(\vartheta)\,d\vartheta + \int_{\vartheta_{MAX}}^{\pi} I_P(\vartheta)\sin(\vartheta)\,d\vartheta = \frac{\sigma_E}{2\pi}(1+m)\,k\,\rho$$

After performing the second integration and calling

9) $$\sigma_{eff} = \left[\sigma_E(1+m) - \frac{1}{2}\left(\cos\left(\vartheta_{MAX}\right)+1\right)\left(\sigma_B + m\,\sigma_E\right) + \Delta\sigma\left(1 - \sin^4\left(\frac{\vartheta_{MAX}}{2}\right)\right) \right]$$

we obtain the value of the normalization constant k which should be characteristic of the experimental configuration but independent from sample to sample:

10) $$k = \frac{2\pi \int_0^{\vartheta_{MAX}} I(\vartheta)\sin(\vartheta)\,d\vartheta}{\sigma_{eff}}$$

The desiderata differential cross section is then:

11) $$\frac{d\sigma}{d\Omega} = \frac{I(\vartheta)}{k\,\rho} - \frac{m\,\sigma_E}{4\pi} - \frac{P(\vartheta)}{k\,\rho}$$

3. RESULTS

As an example of applicability of this method of data reduction we show the results which we obtained in the case in the case of deuterated liquid benzene [6] at standard thermodynamic conditions. The measurements have been taken at the TRIGA reactor (1 MW) of the ENEA-Casaccia research center. We used two different neutron wavelengths (1.2 and 0.9 Å), obtained by a (111) Ge reflection, to cover a wider range of transferred momentum. The scattered intensity was measured in the range from 7.2° to 120.2° stepping 1° corresponding to an accessible Q range included between 0.62 Å^{-1} and 12.0 Å^{-1}. The sample (100% of deuteration) was contained at room temperature inside a vanadium can 5.0 cm high, 1.18 cm in internal diameter and with 0.01 cm of wall thickness.

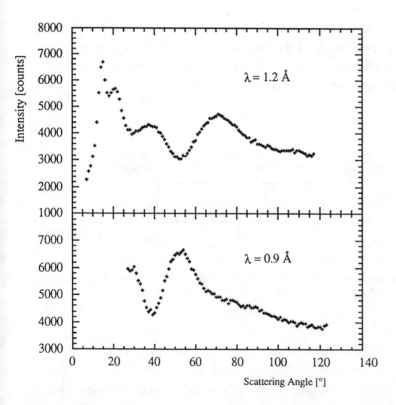

Figure 1. Raw diffraction data liquid deuterated benzene for the two wavelengths used in the experiment as a function of the scattering angle. The data have already been corrected for background and self absorpiton.

In figure 1 we show the raw data as a function of the scattering angle ϑ corrected for background, geometry and absorption. The experimental scattering cross section was determined by a transmission measurement. This method allows to determine the total cross section σ_T which is the sum of the absorption and scattering contributions. In the case of liquid deuterated benzene the absorption neutron cross section, σ_A, is practically equal to 0 ($\sigma_A <$ 0.03 barns) and its contribution to σ_T can be neglected.

The transmission measurements have been performed at the two incident neutron wavelengths used in the experiment by measuring the intensity of a (111) reflection of a copper single crystal. The Cu crystal was at the sample position while the deuterated benzene sample contained into a aluminium flat vessel 1 cm thick was put between the Cu crystal and the monochromator. The attenuation of the reflection is proportional to the total cross section of the sample. We choose this method to measure the transmission to minimize the contribution of the small angle scattering.

Once the experimental scattering cross section is known we can apply all the relations described in paragraph 2. In table I we report the values of σ_B and σ_E, of the absorption coefficient m and of the constant B of the approximated Placzeck corrections (equation 4b) for the two neutron incident wavelengths used in the experiment.

λ [Å]	σ_B [barns]	σ_E [barns]	μ [cm^{-1}]	B [barns Å^{-2}]
1.2	79.16	60.7	0.412	0.0267
0.9	79.16	54.1	0.367	0.0205

Table I. Values of the parameters employed for the data reduction for the two wavelengths used in the xperiment. For maior details please refer to the text.

The normalized data are shown in figure 2 together with the calculated molecular structure factor. In this case the data reduction described completely succeeded. We have to stress that we did not have to apply any further corrections to the data to obtain this kind of agreement. Moreover the measurement of the the experimental scattering cross section σ_E is crucial for all the data reduction and an error smaller than 0.2 % was necessary to correct exactly the inelastic contribution. Before a long shutdown of the TRIGA reactor we have verified the goodness of this data reduction method with other liquids as CCL_4 and concentrated ionic solutions in heavy water. On the other hand we were not able to reduce scattering data taken from liquid water. One of the most serious problem in this case is the expression for the inelastic component which heavily determines all the data reduction procedure. Work is still in progress in this subject.

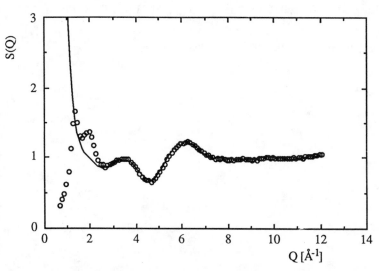

Figure 2. Static structure factor of deuterated liquid benzene at standard thermodynamic conditions. The line is the molecular structure factor obtained by using 1.38 Å for the C-C bond length and 1.09 Å for the D-C bond length.

4. CONCLUSIONS

We have oulined a self consistent normalization procedure for scattering data taken from liquids sample. We have checked the applicability of this method in several occasions getting very incouraging results. The advantage of this method lies on the opportunity of avoiding a normalization run with all the problems which it concerns. On the other hand we could not get an accurate data normalization in the case of heavy water. For the next future we plan to perform more accurate measurements in different kinds of systems to check the general validity of the outlined procedure. We are also working to determine an analogous data reduction method for TOF instruments.

References

[1] Enderby, J.E., and Neilson, G.W., 1981, *Rep. Progr. Phys.*, **44**, 593

[2] Enderby, J.E., and Neilson, G.W., 1979, in *Water, a Comprehensive Traetise*, vol. 6, ed. Franks, Plenum, New York

[3] Placzeck, G., 1952, *Phys. Rev.*, **86**, 377

[4] Powles, J.G., 1979, *Mol. Phys.*, **37**, 623

[5] Yarnell, J.L., Katz, M.J., Wenzel, R.G., and Koemig, S.H., 1973, *Phys. Rev.* a, **7**, 2130

[6] Felici, R, Cilloco, F., and Bosi, P., 1990, *Mol. Phys.*, in press

Inst. Phys. Conf. Ser. No 107: Chapter 3
Paper presented at Neutron Scatt. Data Anal. Conference, Rutherford Appleton, 1990

ATLAS: A suite of programs for the analysis of time-of-flight neutron diffraction data from liquid and amorphous samples

A.C.Hannon, W.S.Howells and A.K.Soper

ISIS, R3, Rutherford Appleton Laboratory, Chilton, Didcot, Oxon OX11 0QX, UK.

Abstract. A description is given of a suite of computer programs recently developed at ISIS for the analysis of time-of-flight neutron diffraction data from non-crystalline samples. The various stages of the analysis are discussed, giving particular emphasis to those aspects in which the analysis of time-of-flight diffraction data differs from the analysis of diffraction data from a constant wavelength source.

1. INTRODUCTION

The ATLAS suite of programs (Soper *et al* 1989) enables experimentalists to analyse non-crystalline time-of-flight (t-o-f) neutron diffraction data speedily and correctly. This suite was conceived towards the end of 1988 with the initial purpose of providing a means by which users of the LAD diffractometer (Howells 1980) at ISIS could analyse their data. Some of the early users of this diffractometer had experienced problems with data analysis due to a lack of appreciation of some of the differences between the analysis of t-o-f diffraction data and reactor (ie constant wavelength) diffraction data. These differences arise as a consequence of the dispersive nature of the t-o-f technique. A workshop was held in April 1989 to introduce the ATLAS suite to potential users and since then it has been used regularly for the successful analysis of LAD data. The suite is written in such a way that it can be applied to data taken on other t-o-f diffractometers with ease and it has now started to be used for data taken on the new SANDALS diffractometer (Soper 1989). The ATLAS suite is based around the use of the GENIE package (David *et al* 1986) in order to make it more accessible to users and to allow for easy display and manipulation of the data at any stage. The suite is written as a number of separate units to allow for maximum flexibility and to enable each step of the analysis to be evaluated critically. Thus if at a particular stage the user prefers some approach other than that adopted by the relevant ATLAS unit, it is possible to perform that stage by the preferred method and afterwards to continue using the ATLAS suite. A manual for the ATLAS suite (Soper *et al* 1989) is available from the authors of this paper.

2. OUTLINE OF NEUTRON DIFFRACTION THEORY

2.1. General Theory

The quantity measured in a neutron diffraction experiment is the differential cross-section. This is defined as;

$$\frac{d\sigma}{d\Omega} = \frac{\left(\begin{array}{l}\text{Number of neutrons of wavelength } \lambda \text{ scattered per unit}\\ \text{time into the small solid angle } d\Omega \text{ at scattering angle } 2\theta\end{array}\right)}{N\Phi(\lambda)d\Omega} = I(Q) \qquad (1)$$

where N is the number of scattering units in the sample and $\Phi(\lambda)$ is the incident neutron flux at wavelength λ. Q is the magnitude of the scattering vector (momentum transfer) for elastic scattering (ie. assuming no change of neutron energy in the scattering process), given by;

$$Q = \frac{4\pi \sin\theta}{\lambda} \qquad (2)$$

The differential cross-section may be expressed as a sum of two terms (Placzek 1952);

$$I(Q) = I^S(Q) + i(Q) \qquad (3)$$

where the first term is known as the self scattering and the second term is known as the distinct scattering. For non-crystalline samples this formulation of the equations is more convenient than the alternative but equivalent formulation in terms of coherent and incoherent scattering. The self and distinct scattering are given by;

$$I^S(Q) = \sum_l \frac{n_l \sigma_l^{scatt}}{4\pi}(1 + P_l(Q,\theta))$$

$$Qi(Q) = \int_0^\infty D(r)\sin(Qr)dr \qquad (4)$$

where the l summation is taken over elements in the sample, n_l is the number of l atoms in the scattering unit and σ_l^{scatt} is the total bound atom scattering cross-section for element l. $P_l(Q,\theta)$ represents a correction for the effects of inelasticity. It corrects for the fact that a real experiment can never fulfill the theoretical requirement that the inelastic scattering be integrated at constant detector efficiency along a line of constant Q in Q-ω-space. Although written for convenience as a function of only Q and θ, the correction is actually a function of many experimental variables such as the temperature T of the sample. To first approximation the inelasticity correction for the distinct scattering $i(Q)$ is zero. In the static approximation $P_l(Q,\theta)$ is ignored. $D(r)$ is the so-called differential correlation function (a formal definition is given below). By Fourier transformation;

$$D(r) = \frac{2}{\pi}\int_0^\infty Qi(Q)\sin(rQ)dQ \qquad (5)$$

Thus the essential stages in the analysis of a diffraction experiment on a non-crystalline sample are;
• measure and correct $I(Q)$,
• extract $i(Q)$ by subtracting $I^S(Q)$,
• derive $D(r)$ by Fourier transformation of the interference function $Qi(Q)$.

The partial pair distribution function $g_{ll'}(\mathbf{r})$ is defined so that $g_{ll'}(\mathbf{r})d\mathbf{r}$ is the average number of l' atoms in a volume $d\mathbf{r}$ which is $\mathbf{r}$ away from an l atom. A formal definition of this statement is as follows;

$$g_{ll'}(\mathbf{r}) = \frac{1}{N_l} \sum_{\substack{j=1 \\ j \neq j'}}^{N_l} \sum_{j'=1}^{N_{l'}} \langle \delta(\mathbf{r} + \mathbf{R}_j(0) - \mathbf{R}_{j'}(0)) \rangle \qquad (= G_{ll'}^D(\mathbf{r}, 0)) \qquad (6)$$

where the angular brackets denote a thermal average at the temperature T of the sample and $\mathbf{R}_j(t)$ represents the position of atom j at time t. The summations j and j' are taken respectively over the N_l or $N_{l'}$ atoms of elements l and l', excluding possible terms where j and j' refer to the same atom. ($G_{ll'}^D(\mathbf{r}, t)$ is a Van Hove (1954) correlation function.) Non-crystalline samples are usually isotropic with the result that $g_{ll'}(\mathbf{r})$ depends only on the magnitude r of $\mathbf{r}$ and not on its direction. Thus $g_{ll'}(\mathbf{r})$ may be replaced by $g_{ll'}(r)$. Partial correlation functions and partial differential correlation functions are related to $g_{ll'}(r)$ as follows;

$$\begin{aligned} t_{ll'}(r) &= 4\pi r g_{ll'}(r) \\ d_{ll'}(r) &= t_{ll'}(r) - t_{l'}^0(r) \\ t_{l'}^0(r) &= 4\pi r g_{l'}^0 \\ g_l^0 &= \frac{N_l}{V} \end{aligned} \qquad (7)$$

where V is the volume of the sample and g_l^0 is the macroscopic number density of l atoms. Note that the definition of $g_{ll'}(r)$ adopted here differs from that used previously in the ATLAS manual (Soper *et al* 1989) by a factor $g_{l'}^0$. The present definition is more typical of work in the amorphous solids field and results in a $g_{ll'}(r)$ which tends at high r to $g_{l'}^0$, whereas the previous definition is more commonly used by liquids workers and involves a high r limit for $g_{ll'}(r)$ of one. A consequence of the definition used here is that $g_{ll'}(r)$ is not symmetric (ie $g_{ll'}(r) \neq g_{l'l}(r)$);

$$\frac{g_{l'l}(r)}{g_{ll'}(r)} = \frac{g_l^0}{g_{l'}^0} \qquad (8)$$

The total correlation function $T(r)$ and the differential correlation function $D(r)$ are expressed in terms of the partial functions as follows;

$$\begin{aligned} T(r) &= \sum_{ll'} n_l \bar{b}_l \bar{b}_{l'} t_{ll'}(r) \\ D(r) &= \sum_{ll'} n_l \bar{b}_l \bar{b}_{l'} d_{ll'}(r) \\ T(r) &= D(r) + T^0(r) \\ T^0(r) &= 4\pi r g^0 \left(\sum_l n_l \bar{b}_l \right)^2 \end{aligned} \qquad (9)$$

where $\bar{b}_l$ is the coherent scattering length for element l and $g^0 = \frac{N}{V}$ is the macroscopic number density of scattering units. Note that for real-space functions a capital letter indicates a function dependent upon scattering lengths whilst a small letter indicates a function which is independent of scattering lengths. Without isotopic substitution one cannot make a measurement on a polyatomic sample and obtain a function which does not depend upon scattering lengths. This is the reason why in practice one must use functions like $T(r)$ and $D(r)$ which involve scattering properties of the sample as well as structural properties. For solids an important additional reason to prefer the use of functions such as $T(r)$ and $D(r)$ which scale as r, as opposed to functions such as $g(r)$ which scales as a constant or the ubiquitous radial distribution function which scales as r^2, is that in the harmonic approximation they are broadened symmetrically by thermal motions (Wright and Sinclair 1985). This can be of considerable advantage in differentiating between static and thermal disorder (Hannon 1989).

2.2. The Total Cross-section

The total cross-section $\sigma^{tot}(\lambda)$ for the removal of neutrons of wavelength λ from a beam by a sample may be separated into two contributions, due to absorption and scattering respectively;

$$\sigma^{tot}(\lambda) = \sigma^{abs}(\lambda) + \sigma^{scatt}(\lambda) \tag{10}$$

If the neutron energy is not close to that of a resonance for any of the nuclei in the sample then the absorption cross-section $\sigma^{abs}(\lambda)$ is proportional to wavelength;

$$\sigma^{abs}(\lambda) = \lambda \sigma^{abs}(\lambda = 1\text{Å}) \tag{11}$$

If the neutron energy is close to that of a resonance then the absorption cross-section is given by the much more complicated Breit-Wigner expression.

The total scattering cross-section may be obtained by integrating the single differential cross-section over all solid angles;

$$\sigma^{scatt}(\lambda) = \int_0^{4\pi} \frac{d\sigma}{d\Omega} d\Omega \tag{12}$$

which may be recast as;

$$\sigma^{scatt}(\lambda) = \frac{8\pi}{Q_m^2} \int_0^{Q_m} Q I(Q) dQ \tag{13}$$

where $I(Q)$ is defined by eqn (1) and Q_m is the maximum accessible value of Q (corresponding to a scattering angle $2\theta = 180°$);

$$Q_m = \frac{4\pi}{\lambda} \tag{14}$$

As indicated by eqn (4) the diffraction pattern $I(Q)$ of a sample exhibits features related to the atomic structure. Eqn (13) shows that $\sigma^{scatt}(\lambda)$ must also exhibit features related to the atomic structure. In the case of crystalline samples this leads to the phenomenon of Bragg edges in the measured transmission cross-section. For non-crystalline samples the peaks and troughs of $\sigma^{scatt}(\lambda)$ correspond to the points of greatest slope of $I(Q)$ and vice versa. At low wavelength $\sigma^{scatt}(\lambda)$ becomes constant tending to the free atom scattering cross-section whilst at high wavelength $\sigma^{scatt}(\lambda)$

tends to a constant value which is less than the bound atom scattering cross-section by a factor depending on the compressibility limit ($Q \to 0$) of $I(Q)$. The total cross-section $\sigma^{tot}(\lambda)$ may only be evaluated directly from the tabulated absorption and scattering cross-sections for a sample which has a coherent scattering cross-section of zero; this is nearly the case for vanadium. Note that eqn (13) is only strictly correct in the static approximation. If the effects of inelasticity are included then $\sigma^{scatt}(\lambda)$ is given by the integral of the differential cross-section measured in a constant wavelength diffraction experiment rather than that measured in a t-o-f experiment (Hannon 1990).

2.3. Termination Of The Fourier Integral.

In practice it is only possible to measure the diffraction pattern up to some finite momentum transfer Q_{max}, and not to infinity. Hence for real experimental data the integral of eqn (5) cannot be performed with the given limits, but only with the upper limit replaced by Q_{max}. This is equivalent to multiplying the cross-section by a modification function $M(Q)$ which is a step function cutting off at $Q = Q_{max}$. The resultant correlation function is;

$$D'(r) = \frac{2}{\pi} \int_0^\infty Qi(Q)M(Q)\sin(rQ)\mathrm{d}Q \tag{15}$$

Applying the convolution theorem gives;

$$D'(r) = \int_0^\infty D(r')(P(r-r') - P(r+r'))\mathrm{d}r' \tag{16}$$

where r' is a dummy variable and;

$$P(r) = \frac{1}{\pi} \int_0^\infty M(Q)\cos(rQ)\mathrm{d}Q \tag{17}$$

With $M(Q)$ equal to a step function the real-space peak function $P(r)$ has a strong oscillatory component which extends over quite a large range of r on either side of the main peak. This leads to spurious features in the correlation function, known as termination ripples. An approach frequently used to reduce the termination ripples involves the use of some sort of damping function for the modification function $M(Q)$ (Waser and Schomaker 1953). The most widely used modification function is that due to Lorch (1969);

$$M(Q) = \frac{\sin(Q\Delta r)}{Q\Delta r} \quad \text{for} \quad Q < Q_{max}$$
$$= 0 \quad \quad \text{for} \quad Q > Q_{max} \tag{18}$$

where;

$$\Delta r = \frac{\pi}{Q_{max}}$$

Use of the Lorch modification function results in a peak function of height $0.1876Q_{max}$ and with a FWHM of $\frac{5.437}{Q_{max}}$. The effect of the Lorch function is to greatly reduce termination ripples, although at the expense of some real space resolution (the step function modification function has FWHM $\frac{3.8}{Q_{max}}$). The real space resolution of a measured correlation function is determined by Q_{max} and thus a diffraction experiment should be performed to as high a value of Q_{max} as possible. Whilst for a constant wavelength diffraction experiment Q_{max} is rigidly limited by the maximum possible scattering angle, for a t-o-f diffraction experiment Q_{max} is only limited by the counting time required to achieve adequate statistics at high Q.

3. THE TIME-OF-FLIGHT NEUTRON DIFFRACTION TECHNIQUE

The t-o-f technique involves the production of a pulse of neutrons with a wide range of energies. In a t-o-f diffraction experiment the neutrons travel a distance L along the incident flight path, are scattered by the sample through an angle 2θ and then travel a distance L' along the scattered flight path to the detector. As they travel the total flight path $L + L'$ neutrons of different energy become separated. Thus a measurement of the t-o-f of a neutron from source to detector enables its wavelength to be determined. In a t-o-f diffraction experiment the momentum transfer Q is varied by maintaining the detector at fixed scattering angle and allowing the neutron wavelength to vary. This is to be contrasted with a diffraction experiment using a continuous source of neutrons where Q is varied by maintaining a fixed neutron wavelength and varying the angle of the detector.

The dispersive nature of the t-o-f neutron diffraction technique has a number of important consequences for the analysis of data;

- The data must be normalised according to the flux shape of the incident neutron pulse as shown by the denominator $\Phi(\lambda)$ in eqn (1). (The general behaviour of $\Phi(\lambda)$ is apparent in figure 3a below.)

- It is especially important that t-o-f data be corrected for detector deadtime since the count rate varies to a much greater extent over a raw spectrum than for the continuous source case. (The deadtime correction corrects for the fact that after detecting a neutron a detector is unable to detect another neutron for a period of time.)

- In the calculation of the corrections for attenuation and multiple scattering it is necessary to include the wavelength dependence of the total cross-section $\sigma^{tot}(\lambda)$. Whilst it is widely known that the absorption cross-section varies with wavelength, the wavelength variation of the scattering cross-section due to atomic structure and to inelasticity is not so well known, and all such wavelength dependences must be taken into account.

- The occurrence of nuclear resonances can sometimes pose a problem for t-o-f diffraction. However, the range of Q over which the data are affected by the resonance is different for detectors at different angles (the resonance occurs at constant energy). Thus, provided that data are available from detectors at more than one appreciably different angle, this problem may be overcome in analysing the data by combining data from detectors at different angles. Regardless of the occurrence of nuclear resonances, it is usual in the analysis of non-crystalline t-o-f diffraction data to combine data from different angles in order to extend the Q-range of the final result and to improve the statistics.

- In order to evaluate the inelasticity correction for a t-o-f diffraction experiment it is necessary to consider the form of the correction with Q varying and with 2θ fixed (ie for a stationary detector). The form of the correction is such that the measured self scattering $I^S(Q)$ is slightly below the static approximation value at high Q, whilst at quite low Q it is found that $I^S(Q)$ begins to rise markedly above this value (see figure 6 below). The Q-dependence of the t-o-f inelasticity correction is thus very different from that of the constant wavelength correction; in a constant wavelength experiment $I^S(Q)$ is only slightly below the static approximation value at low Q, whilst at high Q it steadily falls away from this value. The t-o-f correction is somewhat more complicated than the constant wavelength correction in that it has additional dependences upon the flight path ratio $\frac{L}{L'}$ and incident spectrum shape.

- Whilst the Q-dependences of the inelasticity corrections for the two types of diffraction experiment are very different, the angle-dependences are qualitatively the same; in both cases $I^S(Q)$ decreases as 2θ is increased. For the analysis of t-o-f data the dependence upon angle has the consequence that detectors at different angles must be analysed separately until the inelasticity correction has been performed (ie until an advanced stage of the analysis).

4. THE USE OF THE ATLAS SUITE

In order to illustrate the use of the ATLAS suite we will show the use of each routine for the same sample so that a simple analysis can be followed through from raw data to final result. The sample chosen for this purpose is a rod of vitreous GeO_2 and the data concerned were taken on LAD (Umesaki *et al* 1988). Figure 1 shows the conceptual sequence of operations involved in the use of the ATLAS suite. Figure 2 is a flowchart giving the full details of all the ATLAS operations to be performed on a glass rod sample; data taken for a glass rod are the simplest to analyse because the effects of a sample container (can) do not need to be taken into account. However, it must be emphasised that the suite enables corrections to be performed for a sample in a can or for a sample in a can inside sample environment equipment such as a furnace.

Since the example data were measured on LAD it is desirable to give a brief description of this diffractometer. The incident neutron beam used on LAD originates in a methane moderator at the ISIS pulsed neutron source. The sample is placed at a distance 10m from the moderator, with a scattered flight path from sample to detectors of approximately 1m. The detectors lie in the horizontal plane with identical arrangements on the two sides. There are ^{3}He detector banks at nominal scattering angles of 5°, 10° and 150° and lithium glass scintillator detector banks at 20°, 35°, 60° and 90°. There is also a monitor detector in the incident beam to measure the neutron flux and a monitor detector in the transmitted beam to measure the sample cross-section. The monitors are a special type of scintillator detector consisting of an array of glass scintillator beads.

- The program NORM performs the initial adding together of runs on the same sample and reduction of data. Firstly the data are corrected for deadtime. For scintillator detectors there is a slight complication in that, because of the way the encoder electronics works, a single scintillator counting a neutron causes the a whole bank to cease counting for a period of time. This is taken into account when performing the

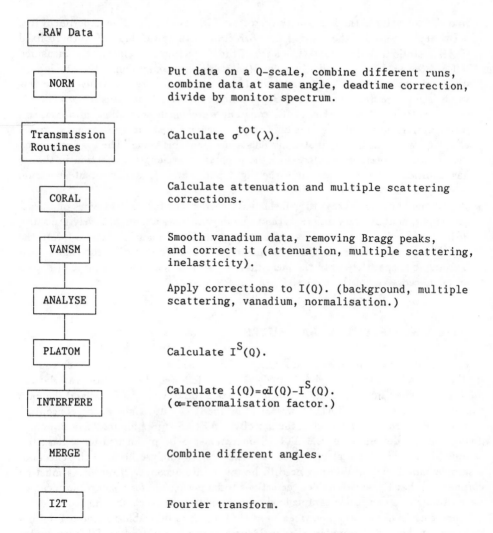

Figure 1. The sequence of operations to be performed when using the ATLAS suite.

correction. Figure 3 shows the effect of the deadtime correction for the example data. The effect is most severe for the monitors since these have the highest count rate. Whilst the effect is very slight for the spectrum shown in figure 3b it should be pointed out that at the time these data were taken the count rate at ISIS was substantially less than it is now; the magnitude of the deadtime correction is now correspondingly larger. The data are normalised to $\Phi(\lambda)$ by dividing the spectra by the incident beam monitor spectrum, after a conversion to a wavelength scale. This causes a complication in that a factor equal to the ratio of the detector and monitor efficiences is introduced. (The effect is apparent in the data shown below in figure 5.) The reason for doing this rather than just making use of the vanadium data to take out the flux shape is that

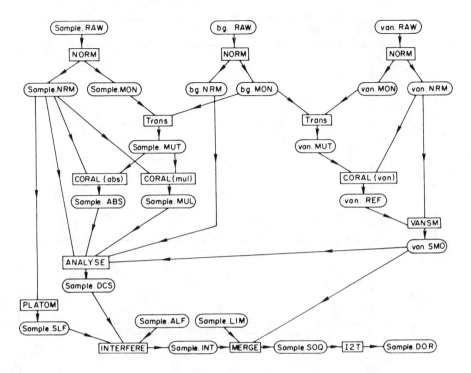

Figure 2. The full details of all the ATLAS operations to be performed in the analysis of an experiment on a sample without a can. Rectangular boxes represent programs whilst boxes with rounded ends represent data files.

this approach removes any change in the flux shape which may have occurred due to changes in moderator temperature. However, there is the possible disadvantage that the deadtime correction is much larger for the monitors than for the other detectors. A reduction of the data is achieved by converting the spectra to a Q-scale and then combining spectra from detector elements at very similar angles into a single spectrum.

• The total cross-section of the sample as a function of wavelength may be evaluated by use of the spectrum measured by the transmission monitor. Two measurements must be made, one with the sample in the beam, one with the sample (but not the can if used) removed. The ATLAS suite includes programs which extract $\sigma^{tot}(\lambda)$ from these measurements either for a flat plate sample which covers the whole neutron beam or for a cylindrical sample. For the flat plate length ma y be evaluated by use of the spectrum measured by the transmission monitor. Two measurements must be made, one with the sample in the beam, one with the sample (but not the can if used) removed. The ATLAS suite includes programs which extract $\sigma^{tot}(\lambda)$ from these measurements either for a flat plate sample which covers the whole neutron beam or for a cylindrical sample. For the flat plate measured total cross-section of GeO_2 corresponds to the second major peak of $I(Q)$ at about 4.8Å^{-1}. This peak in $I(Q)$

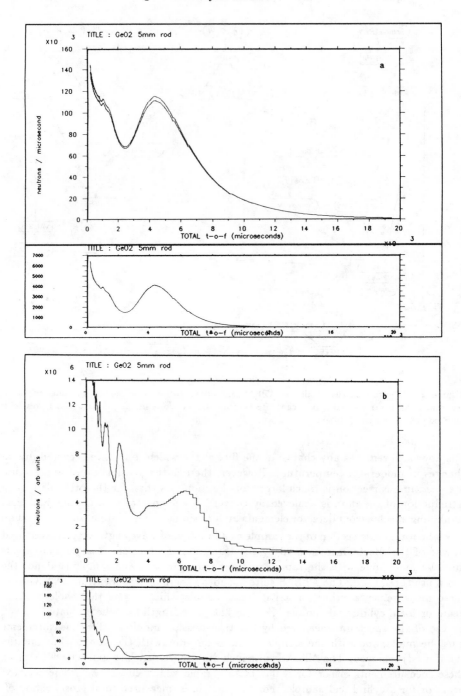

Figure 3. The effect of the deadtime correction for the data on GeO_2. a) for the incident beam monitor b) for one of the two 35° scintillator detector banks. In each case the upper frame shows the t-o-f spectrum before (lower curve) and after (upper curve) the deadtime correction whilst the lower frame shows the difference.

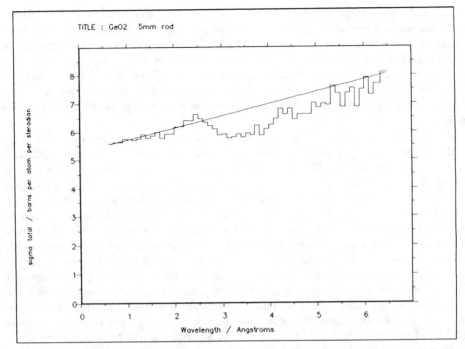

Figure 4. The measured total cross-section of GeO_2 (histogram) together with the total cross-section calculated by ignoring the wavelength-dependence of the scattering cross-section.

corresponds to the negative slope of $\sigma^{tot}(\lambda)$ at a wavelength of about 2.6Å. The peak in $\sigma^{tot}(\lambda)$ at 2.4Å and the trough at 3.2Å correspond to the maximum and minimum slope on either side of the 4.8Å^{-1} peak in $I(Q)$. Note that as is shown by the examples in figure 10 the total cross-section can exhibit considerably more structure than for the example data in figure 4.

• The program CORAL is used to calculate the attenuation and multiple scattering corrections. The calculations may be performed for either a cylindrical or a flat plate sample, with or without a can. For cylindrical geometry a numerical integration method (Soper and Egelstaff 1980) is used which allows corrections involving a furnace or other such sample environment equipment to be evaluated if necessary. The full wavelength dependence of the total cross-section is included in the calculations for both corrections.

• The program VANSM is used to fit the measured spectra for the vanadium calibration sample after a background subtraction. The vanadium data are also corrected using the results of the vanadium calculations by the CORAL program and the results of a standard calculation of the vanadium inelasticity correction which has been performed by the use of the approach of the program PLATOM described below. There are two reasons for fitting the vanadium spectra. Firstly a fit is required in order to remove the Bragg peaks from the spectra: Whilst the coherent scattering length of vanadium is small it is not actually zero with the result that small Bragg peaks are apparent in the vanadium spectra. One of the virtues of t-o-f diffractometers is that they tend to have very good Q-resolution with the consequence that the Bragg peaks of vanadium are clearly prominent and cannot just be ignored. At low Q the Bragg peaks are well separated compared to the resolution width and thus they are given a weight of zero in the fit performed by VANSM. However, at high Q the Bragg peaks merge together and it becomes impossible to weight them out. The second reason for performing a fit is to filter out noise since (excluding the Bragg peaks) vanadium gives rise to smooth slowly varying spectra. A relatively low order Chebyshev polynomial is thus used to perform the fit. A typical fit produced by the program VANSM is shown in figure 5.

• The program ANALYSE is used to calculate the differential cross-section $I(Q)$ of the sample. Firstly a sample minus background subtraction is performed followed by a division by the VANSM fit to the vanadium data and then a subtraction of the CORAL calculation of the multiple scattering from the sample. If a can was used in the experiment then the same three initial operations are performed as for the sample, followed by an application of the attenuation correction for the can as calculated by the CORAL program. A sample minus can subtraction is then performed if necessary. The CORAL calculation of the attenuation correction for the sample is applied and then finally the data are scaled by a factor involving the number of scattering units in the beam to produce an absolute normalisation. For the example GeO_2 experiment figure 5 shows for one detector bank the sample, vanadium (with and without smoothing) and background data prior to the use of ANALYSE and the result obtained afterwards. The differential cross-section $I(Q)$ obtained from the program ANALYSE is that which would be measured in an ideal t-o-f diffraction experiment (ie an experiment without background, can scattering, attenuation or multiple scattering). It should be noted, however, that the result of ANALYSE is necessarily a separate differential cross-section for each detector bank since the inelasticity correction is a function of angle. The purpose of the division by the vanadium data performed by ANALYSE is twofold. Firstly it enables an estimate of the absolute normalisation of $I(Q)$ to be achieved. Secondly it removes the ratio between detector and monitor efficiencies which is introduced by the program NORM.

• The inelasticity correction may be performed by use of the program PLATOM. The output of this program is a calculation of the self scattering $I^S(Q)$ (eqn (4)) for each detector bank. The effect of inelasticity $P(Q,\theta)$ is included by use of the expressions given by Howe *et al* (1989). The approach used to calculate these expressions uses a series expansion involving the moments of $S(Q,\omega)$. This approach was originally proposed by Placzek (1952) and was subsequently developed in a more realistic form for a constant wavelength diffraction experiment by Yarnell *et al* (1973). In the program PLATOM the dependence of the t-o-f correction upon the flux shape has been included by fitting to the incident flux shape and making use of the results of the fit in the program. Figure 6 shows the results of the PLATOM calculation for GeO_2. In the static approximation the self scattering for GeO_2 would have a value of 0.453 barns per steradian per atom. Note that the effects of inelasticity are most severe at low Q and that the inelasticity correction increases with increasing angle.

• The program INTERFERE is used to subtract the calculated self scattering $I^S(Q)$ from the measured differential cross-section $I(Q)$ to yield the distinct scattering $i(Q)$. A separate $i(Q)$ curve is obtained for each detector bank. It is generally found that the normalisation achieved thus far is not perfect with the result that $i(Q)$ does not oscillate precisely about zero as it should (figure 7). One reason for this normalisation discrepancy is simply that the amount of sample in the beam cannot usually be esti-

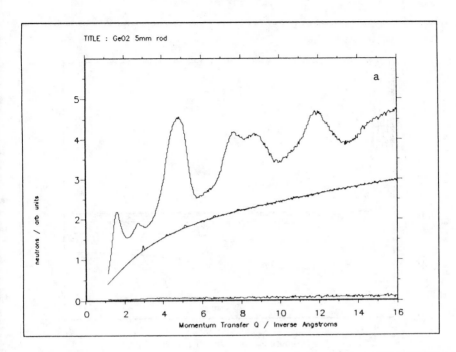

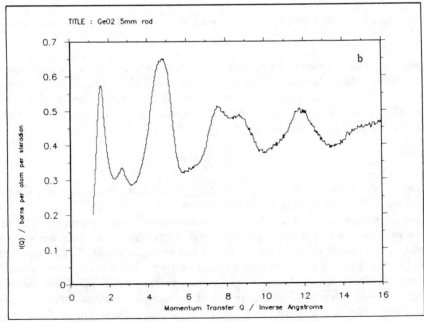

Figure 5. a) The sample (GeO_2), vanadium and background data for the LAD 35° left detector bank are shown prior to the use of ANALYSE. The vanadium data are shown both with and without smoothing. b) The corresponding result from ANALYSE.

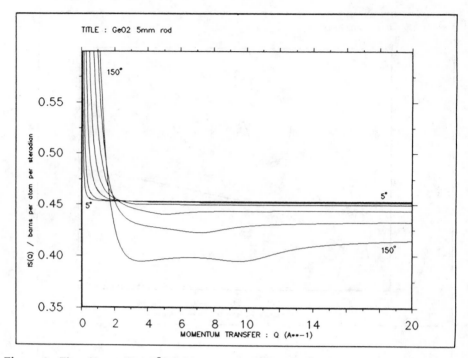

TITLE : GeO2 5mm rod

Figure 6. The self scattering $I^S(Q)$ for GeO_2 at room temperature as calculated by the program PLATOM for each LAD detector angle.

mated exactly. The program INTERFERE allows the normalisation to be corrected by the use of renormalising factors. The renormalising factor α_j for the jth detector bank is defined so that multiplication of the experimental differential cross-section $I_j^e(Q)$ for that bank produces a correctly normalised result. Thus the distinct scattering calculated by INTERFERE for the jth detector bank is;

$$i_j(Q) = \alpha_j I_j^e(Q) - I_j^S(Q) \tag{19}$$

The values used for the α_j are chosen so that $i_j(Q)$ oscillates about zero, but they do not normally differ from unity by more than about 0.05 .

• Once the effects of inelasticity have been removed by using the program INTER-FERE to subtract the self scattering it is possible to combine data from detectors at different angles. This is achieved by the use of the program MERGE. The algorithm employed in the merging process weights each $i_j(Q)$ with the intensity with which it was measured. The weighting function used by the program MERGE is thus obtained from the corrected intensity data of the vanadium sample. Perfect consistency between detectors at different angles is generally not obtained over the whole Q-range over which they overlap. Hence when running MERGE the user must specify the Q-range to be used for each $i_j(Q)$. The Q-limits specified for each detector bank should be chosen giving consideration to the factors affecting the reliability of $i_j(Q)$. If there is a discrepancy over some Q-range between two detector banks then the one with the

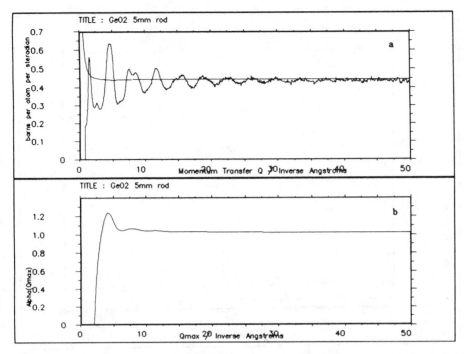

Figure 7. a) The measurement of the differential cross-section $I(Q)$ of GeO$_2$ made by the LAD 60° right detector bank (histogram) together with the calculated self scattering $I^S(Q)$ (smooth line). b) The renormalising factor α calculated for these data as a function of Q_{max}.

smaller inelasticity correction is in general to be preferred. However, other factors such as statistics and Q-resolution may also need to be considered. In practice iterative use of INTERFERE and MERGE may prove necessary to obtain a satisfactory normalisation and final $i(Q)$. Figure 8 shows the final $i(Q)$ for GeO$_2$ obtained as a result of such a procedure. This result appears to be virtually identical to that obtained in a previous constant wavelength measurement (Desa *et al* 1988) over the relatively small Q-range covered in the earlier experiment ($Q_{max} = 23.6\text{Å}^{-1}$).

• Following the calculation of a final combined $i(Q)$ a Fourier transform may be performed if required. The new program I2T performs the transform defined by eqn (15) using either a direct numerical integration method or a fast Fourier transform method. This program is written so that the consistency of the transform with the macroscopic density can be checked. If the normalisation of the data is correct then the slope of the transform at low r should be given by $T^0(r)$ which depends via eqn (9) on the density. This further normalisation check is important if accurate coordination numbers are to be derived. Figure 9 shows some results obtained for the example GeO$_2$ data using the Lorch modification function eqn (18). The $T'(r)$ obtained with a Q_{max} of 23.6Å^{-1} appears to be virtually identical to the result obtained in the previous constant wavelength experiment (Desa *et al* 1988). The increase in real-space resolution arising from the higher Q_{max} of the t-o-f experiment is most apparent in the splitting of the O-O and Ge-Ge peaks at 2.84Å and 3.19Å respectively. The peak which occurs in the

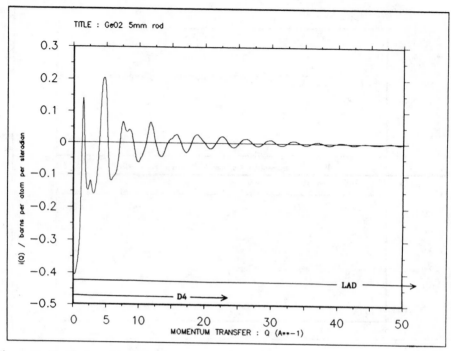

Figure 8. The final combined i(Q) of GeO$_2$.

transform at very low r (of order 0.1Å) arises as a result of imperfect normalisation and thus provides another means by which the accuracy of the normalisation may be evaluated. An additional effect due to increased real-space resolution is to make this 'error peak' sharper.

5. NEW AND FUTURE DEVELOPMENTS

A program has recently been added to the ATLAS suite which calculates the total cross-section of the sample as a function of wavelength using an alternative approach (Hannon 1990) to that discussed in the previous section. Instead of using the measured transmission for the basis of the calculation, $\sigma^{tot}(\lambda)$ is calculated by integrating the diffraction pattern according to eqn (13) and assuming a $\frac{1}{v}$ absorption cross-section (eqn (11)). Figure 10 shows some results obtained using the two approaches. The fall-off of the cross-section of liquid lead at high wavelength (figure 10a) corresponds to the compressibility limit of $I(Q)$. The Bragg edge of Al$_2$O$_3$ at about 4.1Å (figure 10b) corresponds to the intense Bragg peak at approximately 3.0Å^{-1}. The total cross-section calculated by integrating $I(Q)$ has better statistical accuracy than that calculated from the measured transmission and also it avoids possible experimental problems with measuring the transmission. When calculating $\sigma^{tot}(\lambda)$ from the measured transmission for a cylindrical sample using the program described in the previous section the parameters used in the calculation must be adjusted in order to achieve a result with the

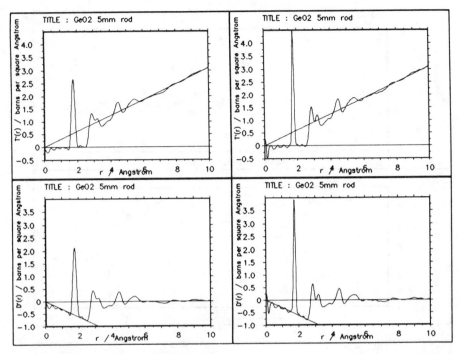

Figure 9. Fourier transforms $T'(r)$ and $D'(r)$ for GeO$_2$. The curves on the left were calculated using $Q_{max} = 23.6 \text{Å}^{-1}$ whilst those on the right have $Q_{max} = 50 \text{Å}^{-1}$. The straight lines correspond to the constant density term $T^0(r)$.

correct absolute normalisation. One reason for this is the discrete way in which the LAD transmission monitor samples the beam. It is not always clear whether the absolute normalisation of the cross-section calculated from the transmission is correct and one use of the new program is to indicate how the correctly normalised cross-section should behave. A disadvantage of the new program is that it tends to be used at an early stage of the analysis, using as the basis for the calculation an $I(Q)$ curve which is not fully corrected; if the corrections are large this can affect the behaviour of the calculated cross-section.

Another new program has recently been added to the ATLAS suite for use in calculating renormalising factors α_j. This program uses the Krogh-Moe (1956) - Norman (1957) method in which taking the low r limit of eqn (15) leads to;

$$\alpha_j = \frac{\int_0^{Q_{max}} Q^2 I_j^S(Q) M(Q) \mathrm{d}Q - 2\pi^2 g^0 (\bar{b}_{av})^2}{\int_0^{Q_{max}} Q^2 I_j^e(Q) M(Q) \mathrm{d}Q} \tag{20}$$

where $\bar{b}_{av}$ is the average coherent scattering length per atom. A result obtained by use of this method for the example GeO$_2$ data is shown in figure 7b. At high Q_{max} the calculated value of α tends to 1.022 - multiplying $I^e(Q)$ by this value causes it to

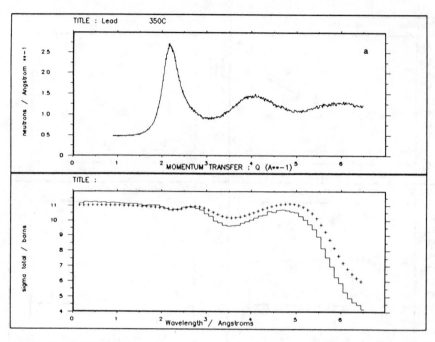

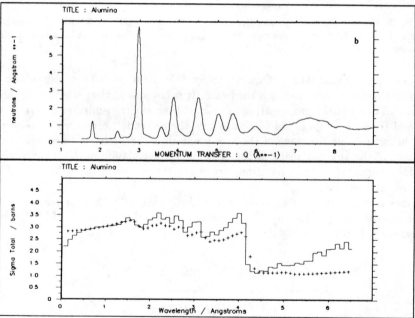

Figure 10. The total cross-section $\sigma^{tot}(\lambda)$ calculated from the measured transmission (histogram) together with that calculated from the diffraction pattern (points) for a) liquid lead (Dahlborg 1989) b) polycrystalline Al_2O_3. In each case the upper frame shows $I(Q)$ whilst the lowere frame shows the total cross-section.

oscillate about $I^S(Q)$. For the method to work well $I^e(Q)$ must cover the full Q-range over which the diffraction pattern is strongly featured. Thus for LAD it works well for all but the highest and lowest angles (150°, 10° and 5°).

One possible future addition to the ATLAS suite would be a program to correct the data for the effects of instrumental Q-resolution (Gardner 1986). This should improve the consistency between data measured at different scattering angles. Another possible future addition to the suite would be to incorporate programs adopting other approaches to the inelasticity correction than that used by the program PLATOM. At present the suite is being adapted to be able to caculate the corrections for flat plate samples at an angle to the beam.

6. CONCLUSION

T-o-f neutron diffraction data for non-crystalline samples may be analysed successfully by the use of the ATLAS suite as described in this paper.

Acknowledgments

We should like to express our gratitude to the attendees of the original ATLAS workshop and many of our subsequent users whose experiences and comments have been of great assistance to us in improving the ATLAS suite.

References

David W I F, Johnson M W, Knowles K J, Moreton-Smith C M, Crosbie G D, Campbell E P, Graham S P and Lyall J S 1986 *Rutherford Appleton Laboratory Report* RAL-86-102
Dahlborg U 1989 *private communication*
Desa J A E, Wright A C and Sinclair R N 1988 *J. Non-cryst Solids* **99** 276
Gardner P P 1986 *Workshop on Neutron Scattering Data Analysis 1986* ed M W Johnson *Inst. Phys. Conf. Ser.* **81** 55
Hannon A C 1989 *PhD Thesis Reading University*
Hannon A C 1990 *to be published*
Howe M A, McGreevy R L and Howells W S 1989 *J. Phys.: Condens. Matter* **1** 3433
Howells W S 1980 *Rutherford Appleton Laboratory Report* RAL-80-017
Krogh-Moe J 1956 *Acta. Cryst.* **9** 951
Lorch E 1969 *J. Phys. C* **2** 229
Norman N 1957 *Acta. Cryst.* **10** 370
Placzek G 1952 *Phys. Rev.* **86** 377
Soper A K and Egelstaff P A 1980 *Nucl. Inst. Meth* **178** 415
Soper A K, 1989 *IOP Conf. Series* **97** 353
Soper A K, Howells W S and Hannon A C 1989 *Rutherford Appleton Laboratory Report* RAL-89-046
Umesaki N, Hannon A C, Wright A C, Brunier T M and Sinclair R N 1988 *unpublished work*
Van Hove L 1954 *Phys. Rev.* **95** 249
Waser J and Schomaker V 1953 *Rev. Mod. Phys.* **25** 671
Wright A C and Sinclair R N 1985 *J. Non-cryst. Solids* **76** 351
Yarnell J L, Katz M J, Wenzel R G and Koenig S H 1973 *Phys. Rev. A* **7** 2130

Inst. Phys. Conf. Ser. No 107: Chapter 4
Paper presented at Neutron Scatt. Data Anal. Conference, Rutherford Appleton, 1990

213

Analysis of neutron reflectivity data using constrained model fitting

J Penfold

Neutron Science Division, Rutherford Appleton Laboratory, Chilton, Didcot, Oxon, OX11 0QX, UK

ABSTRACT: To date neutron reflection data has been analysed predominantly by model fitting, using standard optical equations and methods developed for multilayer optics. These optical methods provide a convenient form for calculating reflectivity profiles, and have been used as a basis for a non-linear least squares refinement. In problems in surface chemistry, isotopic substitution (principally hydrogen–deuterium) can be used to alter refractive index profiles, and this enables additional constraints to be imposed on the model fitting. The application of such methods to the analysis of neutron reflection data, especially the adsorption of surfactants at the air–liquid interface, will be described.

1. INTRODUCTION

The specular reflection of neutrons is now established as a technique for the study of surfaces and interfaces (Penfold et al 1990b), and has been applied particularly to problems in surface chemistry (Penfold et al 1990b) and magnetism (Felcher et al 1987).

The essence of a neutron reflection experiment is to measure the specular reflection as a function of the wave vector transfer, Q, perpendicular to the reflecting surface. This can be related to the neutron refractive index profile normal to the interface, and is often simply related to the scattering length density, yielding information about the composition and density gradient of surfaces and interfaces.

Goldberger and Seitz (1947) have shown that the intensity of the reflected and transmitted neutrons follow the same laws as electromagnetic radiation with the electric vector perpendicular to the plane of incidence. Hence most of the standard formulisms in classical light optics can be used with only minor modifications (Lekner 1987).

The refractive index for neutrons is commonly written as :

$$n = 1 - \lambda^2 A + i\lambda C \tag{1}$$

where, $A = Nb/2\pi$, $C = N\sigma_a/4\pi$, N is the atomic number density, b is the bound coherent scattering length, σ_a is the adsorption cross section, and λ is the neutron wavelength.

In contrast to X-rays, neutron scattering amplitudes vary randomly from element to element, and isotopic substitution can be used to produce large contrasts in the scattering densities. Of particular importance is the large difference in scattering powers of hydrogen and deuterium; this has already been used to great effect in small angle neutron scattering, and will be seen to be of particular importance in the study of surface chemistry.

In surface chemistry the specular reflection of neutrons can be used to provide information about adsorption at interfaces, where not only can the amount adsorbed be determined, but also the structure of the adsorbed layer.   Specifically, hydrogen–deuterium exchange has been used not only to highlight particular parts of the surface, but even to eliminate altogether the reflection from anything other than the adsorbed layer (Penfold et al 1990b).

Although a number of approximate methods exist (Als Nielsen 1985; Crowley et al 1990) for calculating reflectivity profiles, to date the multilayer matrix methods (Born and Wolf 1970; Heavens 1955) have been extensively used. Furthermore, data analysis has in general proceeded by comparison with a model rather than by any attempts at direct inversion.

In this presentation, recent advances in the adaptation of multilayer optical methods to the modelling and interpretation of neutron reflectivities are presented. These optical methods provide a convenient form for calculating reflectivity profiles. They have been used as a basis for a non–linear least squares refinement, and examples of their use to interpret some well defined surfaces and interfaces will be described. In problems in surface chemistry, hydrogen–deuterium substitution can be used to manipulate the refractive index profile at the interface. This enables additional constraints to be imposed upon model fitting, and the application of such methods to the analysis of reflection data from the adsorption of surfactants at the air–solution interface will be discussed.

2. OPTICAL METHODS

The specular reflection at the interface of two bulk media is described by Fresnel's law (Born and Wolf 1970), where for $\theta < \theta_{critical}$ the reflectivity, R, is unity, and for $\theta > \theta_{critical}$

$$R = \left| \frac{n_o \sin \theta_o - n_1 \sin \theta_1}{n_o \sin \theta_o + n_1 \sin \theta_1} \right|^2 \tag{2}$$

where $n_0 n_1$ are the refractive indices of the two media, θ_0 the glancing angle of incidence, and θ_1 the angle of refraction. We can write,

$$n_1 \sin \theta_1 = (n_1^2 - n_o^2 \cos^2 \theta_o)^{1/2} \tag{3}$$

where for $\theta > \theta_{critical}$ $n_1 \sin \theta_1$ is real, for $\theta < \theta_{critical}$ $n_1 \sin \theta_1$ is imaginary (corresponding to an evanescent wave), and for $\theta = \theta_{critical}$ $n_1 \sin \theta_1$ is identically zero.

For a single uniform film at the interface the reflectivity can be written exactly as (Born and Wolf 1970) :

$$R = \left| \frac{r_{01} + r_{12}\, e^{-2i\beta}}{1 + r_{01} r_{12}\, e^{-2i\beta}} \right|^2 \tag{4}$$

where r_{ij} is the Fresnel coefficient at the ij interface such that,

$$r_{ij} = p_i - p_j \,/\, p_i + p_j \tag{5}$$

$p_i = n_i \sin \theta_i$, $\beta = (2\pi/\lambda)n_1 d \sin \theta_1$, d is the film thickness, and the subscripts 0,1,2 refer to the air, film and substrate respectively.

To consider an interface with many discreet layers a general solution must be adopted. Many of the early calculations were based on the standard method described by Born and Wolf (1970). By applying the condition that the wavefunctions and their gradients must be continuous at each boundary, a characteristic matric for each layer can be defined, such that for the jth layer,

$$M_j = \begin{bmatrix} \cos \beta_j & -(i/p_j)\sin \beta_j \\ -ip_j \sin \beta_j & \cos \beta_j \end{bmatrix} \tag{6}$$

The resultant reflectivity for n layers is then given by the elements of the resultant matrix $M_R = [M_1][M_2] ---- [M_n]$, such that,

$$R = \left[\frac{(M_{11} + M_{12}\, p_s)p_a - (M_{21} + M_{22})\, p_s}{(M_{11} + M_{12}\, p_s)p_a + (M_{21} + M_{22})p_s} \right]^2 \tag{7}$$

where the subscripts a and s refer to the air and substrate.

The treatments so far have considered only ideal interfaces, and some consideration must be given to surface and interfacial imperfections. Long range surface undulations contribute to the reflectivity profile in a way similar to beam divergence. However, if the surface is not entirely smooth, its local roughness will modify the specular reflectivity in a manner similar to that of a diffuse interface. Névot and Crocé (1980) showed that for a gaussian distributed height–height correlation function at a bulk interface the reflected intensity will be modified by a factor of the form,

$$I(Q) = I_0(Q) \exp(-q_0 q_1 <\sigma>^2) \tag{8}$$

where $I(Q)$, $I_0(Q)$ are the reflected intensities with and without surface roughness, $<\sigma>$ is the root mean square roughness, $q_0 = 2k\sin\theta_0$, $q_1 = 2k\sin\theta_1$, and k is the neutron wave vector.

Cowley and Ryan (1987) have extended this treatment of interfacial roughness to the case of thin films by applying a similar gaussian factor to the Fresnel coefficients in equation 4 such that,

$$r_{ij} = \left(\frac{p_i - p_j}{p_i + p_j}\right) \exp(-0.5\, q_i q_j <\sigma>) \tag{9}$$

It has been shown (Felici et al 1989) that this approximation is equivalent to using the matrix method (see equations 6 and 7) to describe a gaussian density distribution at an interface by dividing it into a series of discreet layers. This approach to interfacial roughness is not in general convenient, as numerically it can quickly become unwieldly. We have, therefore, chosen to use an alternative multilayer method to calculate reflectivity profiles in order to overcome this problem. An appropriate method is that of Abeles (Heavens 1955) which defines a characteristic matrix per layer in terms of Fresnel coefficients and phase factors, derived in optical terms from the relationship between electric vectors in successive layers. A characteristic matrix per layer (analogous to that defined in equation 6) is then defined as,

$$C_j = \begin{bmatrix} e^{i\beta_{j-1}} & r_j\, e^{i\beta_{j-1}} \\ r_j\, e^{-i\beta_{j-1}} & e^{-i\beta_{j-1}} \end{bmatrix} \tag{10}$$

For n layers the matrix elements M_{11}, M_{21} of the resultant matrix $M_R = [M_1][M_2] -- [M_{n+1}]$ gives the reflectivity,

$$R = M_{21} M_{21}^* / M_{11} M_{11}^* \tag{11}$$

A roughened or diffuse gaussian interface can now be conveniently introduced at each boundary using equation 9. This provides a numerically convenient and closed form for calculating reflectivity profiles exactly (Penfold 1988).

3. UNCONSTRAINED MODEL FITTING

The multilayer matrix method of Abeles (equations 10 and 11) modified for interfacial roughness (equation 9) has been used as the basis for the model fitting of reflection data obtained at the reflectometer, CRISP (Penfold et al 1987), at the ISIS pulsed neutron source. The model fitting uses a non linear least squares routine, VA05A, from the Harwell subroutine library (Hopper 1967), which is a compromise between three different algorithms, Newton Raphson, steepest descent and Marquardt. Standard programs (Penfold et al 1990a) now exist, where the reflectivity can be calculated for any definable layer sequence. Each parameter can be refined or fixed, and includes instrumental resolution, a flat background, an arbitrary scale factor, and surface roughness. In addition there are three parameters per layer (thickness, scattering density and interfacial roughness). Within that framework it is also possible to define repeat sequences for multiple bilayers.

The reflectivity data on CRISP is obtained at a fixed angle of incidence with a pulsed white beam, using time–of–flight to measure wavelength. In this way the reflectivity is measured as a function of Q, perpendicular to the reflecting surface. The resolution in Q, δQ, is dominated by the collimation, $\delta\theta$, and hence resolution effects are therefore included simply by an integration over $\delta\theta$.

Before proceeding to discuss the use of constraints in the model fitting of reflection data, some simple examples are presented which indicate the application of unconstrained model fitting.

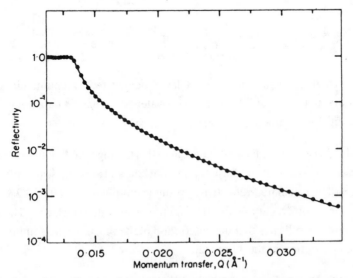

Figure 1. Specular reflection from a $\lambda/10$ optical flat at an angle of incidence $\theta = 0.3°$: • data points, – least squares model fit (for parameters see text).

Figure 1 shows the reflectivity from a bare substrate, a $\lambda/10$ optical flat. The solid line is a model fit for a scattering density of 0.36×10^{-5} Å^2, a surface roughness of 33Å and a $\Delta\theta$ of 4.7%. The quality of the fit indicates that the profile is well described by Fresnel's law, modified for surface roughness and instrumental resolution. For this measurement the intrinsic instrumental resolution $\delta\theta$ is 3.7%; it is assumed that the additional contribution is from surface undulations.

The presence of a thin film on such a substrate will modify the simple Fresnel's reflectivity profile from the bare substrate, and for a well defined layer a series of discrete interference fringes will be observed. The interference pattern from such a well defined layer is shown in Figure 2 for a 48 layer deuterated cadmium eicosanoate Langmuir–Blodgett film deposited onto a polished silicon wafer.

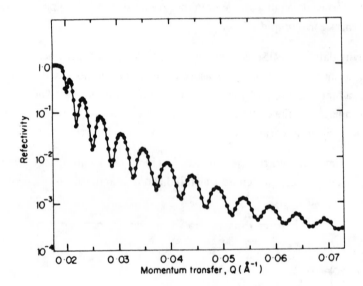

Figure 2. Specular reflection from a 48 layer deuterated cadmium eicosanoate Langmuir–Blodgett film deposited onto a silicon wafer, at $\theta = 0.5°$: • data points, – least squares model fit (for parameters see text).

The solid line is a least squares fit, assuming that in this Q range the Langmuir–Blodgett film can be described as a single uniform layer of thickness 1186Å and scattering density 0.74×10^{-5} Å^{-2}. The calculated reflectivity profile is modified by a $\delta\theta$ of 3.7% and an interfacial roughness of 20Å at the air–film and film–substrate interfaces. The quality of the least squares fit indicates that the description of the film as a single uniform layer is an adequate description in this Q range.

In Figure 3 the neutron reflection from a 20 layer deuterated cadmium docosanoate Langmuir–Blodgett film illustrates an example where a single uniform layer description is inadequate. For values of Q less than 0.11Å^{-1}, the reflectivity is well described by a single uniform layer of thickness 564Å. However, beyond $Q = 0.11 \text{ Å}^{-1}$ a weak Bragg peak arising from the bilayer structure is observed. The calculated curve which now fits the interference fringes and the Bragg peak arises from a model which includes the bilayer structure specifically, with 10 bilayers with a spacing of 57Å ($d_1 = 3.5$Å, $d_2 = 51.5$Å, $Nb_1 = 0.65 \times 10^{-5} \text{ Å}^{-2}$ and $Nb_2 = 0.78 \times 10^{-5} \text{ Å}^{-2}$) and an additional single layer of 27Å at the air–film interface. In both of the fitted curves $\delta\theta$ is found to be 4.5% and the air–film and film–substrate roughnesses are 15Å.

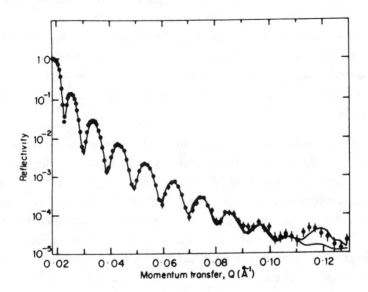

Figure 3. Specular reflection from a 20 layer deuterated cadmium docosanoate Langmuir–Blodgett film deposited onto a silicon wafer at $\theta = 0.5°$: ● data points, – least squares model fit (for parameters see text).

4. CONSTRAINED MODEL FITTING

In the examples presented so far we have used model fitting with no constraints imposed. In the investigation of problems in surface chemistry it is possible to impose constraints upon the model fitting. The first of these results forms the large difference in scattering powers for neutrons between protons and deuterons. Hydrogen–deuterium isotopic substitution can therefore be used in chemical systems to manipulate the refractive index profile. It is particularly important for the adsorption of surfactants at the air–solution interface where isotopic substitution can be applied to both solute and solvent. It is possible to choose the hydrogen–deuterium ratio such that the solvent is null reflecting. If the surfactant is deuterated, then any reflection results entirely from the surface adsorption of the surfactant. It is also possible to eliminate the reflection from the solute and determine the surface profile of the solvent. Isotopic substitution can further be used to highlight particular parts of the solute molecule by selective deuteration. It is then possible to obtain the reflectivity from the same interface viewed with different refractive index profiles, and we can constrain the model of the surface structure to predict each of the reflectivities without adjustment.

These features have been used to some advantage in an early study on the adsorption of the surfactant decyltrimethyl ammonium bromide (Lee et al 1989) at the air–solution interface, where it was possible to determine the detailed surface structure. More recently the same basic model has been used successfully for tetramethyl ammonium dodecyl sulphate (TMDS) (Penfold et al 1989) and sodium dodecyl sulphate (SDS) (Thomas et al 1990) in aqueous solution. In the case of TMDS, by selective deuteration of the tetramethyl ammonium counterion it was also possible to obtain additional information about the degree of counterion binding, and the extent of the diffuse counterion layer.

The simplest basic model that is consistent with the reflectivity data for both the TMDS and the SDS is a two layer model where the first layer adjacent to the vapour phase contains some fraction (1–fhg) of the hydrocarbon chains, and the second layer, adjacent to the aqueous subphase, contains the headgroups, some fraction of the chains (fhg), solvent and bound counterions. The model is then characterised by the three structural parameters, the area per molecule, ahg, fhg, and the extent to which the chain region is fully extended, fc. In the case of the TMDS, an additional parameter is included to account for the degree of counterion binding to the headgroup.

The procedure adopted for analysing the data is to construct the model of the interface and refine the model parameters for one of the refractive index profiles (say deuterated surfactant in null reflecting water). The same model is then used to generate the reflectivity for the different refractive index profiles (obtained form isotopic substitution) for the same system. The model is then only acceptable if it can be used to predict the reflectivity for each refractive index profile with no adjustment of the parameters between the different profiles.

In the detailed specification of the model, it is then possible to impose further constraints determined from chemical information such that for the example illustrated here the detailed model is as follows,

(i) The thickness of the first layer is given by :

$$d_1 = lc \, (1.0 - fhg) . fc \qquad (12)$$

where lc is the length of a fully extended chain.

(ii) The thickness of the second layer is,

$$d_2 = lhg + lc \cdot fhg \qquad (13)$$

where lhg is the extent of the headgroup.

From known molecular volumes and scattering lengths, it is then possible to calculate the scattering densities of the two layers in the model.

In Figure 4 we show the application of this approach to 0.0067M SDS, where we have measured the reflectivity profiles for deuterated SDS in null reflecting water, deuterated SDS in D_2O, and protonated SDS in D_2O.

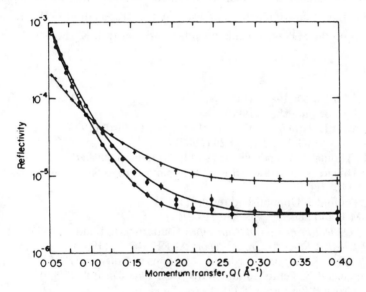

Figure 4. Specular reflection from 0.0067M SDS at $\theta = 0.5°$ for (i) +, deuterated SDS in null reflecting water, (ii) •, deuterated SDS in D_2O, and (iii) o, protonated SDS in D_2O; solid lines are least square model fits (for parameters see text).

The model fits are for an ahg of 43.8Å^2, fhg of 0.2 and fc of 0.87; and show excellent agreement with the data. The significance of these results is considered in detail elsewhere (Thomas et al 1990), as the purpose of this presentation is to describe the methodology and to give some indications as to its effectiveness.

The reflectivity profiles for the different contrasts show particular sensitivity to certain of the model parameters. In the example shown here the reflectivity profile for deuterated SDS in null reflecting water is particularly sensitive to ahg and fc, and it is this reflectivity profile which determined predominantly the density of the surface layer. The reflectivity profile for protonated SDS in D_2O is largely insensitive to fc and ahg, but depends mostly on fhg, whereas the profile for deuterated SDS in D_2O is sensitive to both fhg and fc.

The constrained model fitting has now been applied successfully to a range of surfactant systems which, in addition to the TMDS and SDS, include mixtures of SDS and dodecanol (Thomas et al 1990), fluorocarbon surfactants and mixtures of hydrocarbon and fluorocarbon surfactants (Ottewill et al 1990), and phospholipids (Bayerl et al 1990).

5. ACKNOWLEDGEMENTS

Rob Richardson, Bristol and Thelma Hardman, Reading are acknowledged for providing the Langmuir–Blodgett films used in the presentation. The reflectivity measurements on the SDS were made in collaboration with R K Thomas, Oxford.

6. REFERENCES

Als Neilsen J, *Z Phys* **64** (1985) 411
Bayerl T et al, to be published (1990)
Born M and Wolf E, *Principles of Optics* (Pergamon Oxford) (1970)
Cowley R A and Ryan T W *J Phys D* **20** (1987) 61
Crowley T L, Thomas R K and Willatt A J (1990) to be published
Felcher G P, Hilleke R D, Crawford R K, Haumann J, Kleb R and Ostrowski G, *Rev Sci Inst* **58** (1987) 609
Felici R and Penfold J, Unpublished results (1989)
Goldberger M L and Seitz F, *Phys Rev* **71** (1947) 294
Heavens O S *Optical Properties of Thin Films* (Butterworth, London) (1955)
Hopper M J *UKAEA ResearchGroup Report* AERE–R6912 (1967)
Lekner J, *Theory of Reflection* (Dordrecht, Martinus Nijhoff) (1987)
Lee E M, Thomas R K, Penfold J and Ward R C *J Phys Chem* **93** (1989) 38
Névot L and Crocé P *Phys Appl* **20** (1980) 61
Ottewill R et al, to be published (1990)
Penfold J *Rutherford Appleton Laboratory Report* RAL–88–088 (1988)
Penfold J, Herdman J and Shackleton C, to be published (1990)
Penfold J, Lee E M and Thomas R K *Mol Phys* **68** (1989) 33
Penfold J, Thomas R K, *J Phys: Condensed Matter* **2** (1990) 1369
Penfold J, Ward R C and Williams W G *J Phys E: Sci Inst* **20** (1987) 1411
Thomas R K, Rennie A, Penfold J, Lee E M and Simister E, to be published (1990)

Maximum entropy analysis of neutron reflectivity data—some preliminary results

D. S. Sivia, W. A. Hamilton and G. S. Smith
Manuel Lujan Jr. Neutron Scattering Center
Los Alamos National Laboratory
Los Alamos, NM 87545, U.S.A.

ABSTRACT: The analysis of neutron reflectivity data is akin to the notorious phaseless Fourier problem, well-known in many fields such as crystallography. It is a difficult problem because there are many, very different, nuclear scattering-length density profiles which fit the data. We show the results of analysing simulated reflectivity data, generated from test profiles, using maximum entropy. The results are encouraging, but they also illustrate some of the dangers of the underlying phaseless problem. We indicate how additional information, such as complementary x-ray reflectivity data, can be used to alleviate some of the difficulties.

1. Introduction

A wide variety of interfaces of scientific and technological interest exhibit depth profile structure on length scales of tens to thousands of Ångstrom. Examples include: model biological lipid membranes, analogous to the walls of living cells; semi-conductor multilayers of importance to the electronics industry; organic thin-film chemical sensors; conjugated polymers for electro-optic devices; and thin-film superconductors.

This size range is an order of magnitude less than is accessible by conventional optical techniques. One method of probing interfacial structures on these length scales is to measure the interference effects of the depth profile on the reflection of short wavelength radiation. This extension of visible range optical techniques to shorter wavelengths has been vigorously pursued over the last few years in x-ray reflectometry, and also by the increasingly popular complementary technique of neutron reflectometry. This is possible because thermal neutrons and x-rays exhibit reflection, refraction and interference phenomena analogous to those familiar in classical optics over the desired range of length scales. This analogy can be carried to the definition of a neutron refractive index which is simply related to the local average of the coherent neutron scattering-length density of the nuclei in a material.

Although x-ray sources are generally more intense, neutron reflectometry has several advantages which make it very powerful in its own right. Due to their magnetic interaction, neutrons have a spin dependent refractive index and so the reflection of polarized neutrons is a sensitive probe of surface magnetism. In addition, the neutron scattering-lengths (unlike x-rays) bear no monotonic relationship to the atomic number and, for a given system, the neutron contrast may be much stronger than than that for x-rays. Further, since the scattering-lengths also vary between isotopes of the same element, isotopic substitution may be used to enhance the contrast of a interfacial structure without altering its chemical properties. For example, the vastly different neutron scattering powers of hydrogen and deuterium are widely used to great advantage in highlighting specific sections of organic interface structures.

In Section 2 we consider reflectivity from a data analysis point-of-view and indicate some of the anticipated difficulties by analogy with the closely related, and well-known, phaseless Fourier problem. In Section 3 we briefly review the standard model-fitting approach to the problem and then go on to consider the more ambitious enterprise of finding free-form solutions for the density profile. In Section 4 we show the results of analysing simulated reflectivity data, generated from test profiles, using maximum entropy (MaxEnt). These examples are very instructive, showing encouraging results but also illustrating some of the dangers and difficulties inherent in the reflectivity problem. We show how additional information, in the form of complementary x-ray reflection data, can be very useful, and we conclude with Section 5.

2. The problem

In neutron reflectometry we want to infer the density profile $\beta(z)$, where β is the average neutron scattering-length density and z is the depth below the surface of the sample, given measurements of the reflectivity R as a function of the scattering vector Q (see Fig.1). This inference can be summarised with the conditional probability distribution $\mathrm{prob}[\beta(z)|R(Q)]$, where "|" means "given"; we implicitly assume, as given, knowledge about the experimental setup. Our best estimate of the density profile, in the light of the data, is given by that $\beta(z)$ which maximises this probability distribution function (p.d.f.).

To compute this p.d.f., we need to use Bayes' theorem (see, for example, Jeffreys 1939, or Jaynes 1986). Bayes' theorem relates the p.d.f. we require to two others, one of which can be computed and the other "guessed"; it states that:

$$\mathrm{prob}[\beta(z)|R(Q)] \propto \mathrm{prob}[R(Q)|\beta(z)] \times \mathrm{prob}[\beta(z)] .$$

The term on the far right, $\mathrm{prob}[\beta(z)]$, is called the *prior* p.d.f. and represents our state-of-knowledge (or ignorance) about $\beta(z)$ before we have any data. Our prior state-of-knowledge is modified by the data through the so-called *likelihood function*, $\mathrm{prob}[R(Q)|\beta(z)]$, which tells us how likely it is that we would have obtained our particular data-set if we were given a (trial) density profile. The likelihood function is often (approximated to be) of the form $\exp(-\chi^2/2)$, where χ^2 is the usual sum-of-squared-residuals misfit statistic. The product of the prior p.d.f. and the likelihood function yields the *posterior* p.d.f. we require and represents our state-of-knowledge about the density profile after we have obtained the data.

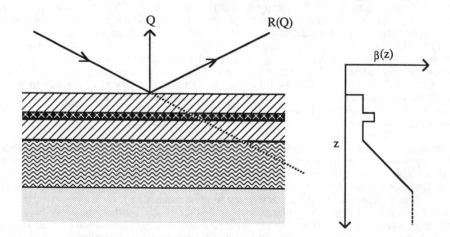

Figure 1: A schematic illustration of a reflectivity experiment. We wish to infer the depth profile $\beta(z)$, given measurements of the reflectivity R as a function of the scattering vector Q.

The prior p.d.f. is often uniform (flat), or at least well-behaved in the sense of having a single maximum (e.g. an entropic prior). The likelihood function for the reflectivity problem, on the other hand, can be very badly-behaved having many (local) maxima (see Fig. 2); this is because the data are related to $\beta(z)$ through a non-linear transformation (Lekner 1987, Felcher 1988, Russell 1990). The posterior p.d.f., which we seek to maximise, also inherits much of the complex topology of the likelihood function and this is the major source of difficulty in the analysis of reflectivity data.

We can anticipate some of the problems in analysing reflectivity data by noting that this problem is very similar to the phaseless Fourier problem, well-known in crystallography (for example). In the limit of small reflectivity, $R<<1$, and large scattering vector, $Q>>0$, the data (with perfect instrumentation) are related to the derivative of the density profile $d\beta/dz$ by a phaseless Fourier transform:

$$R(Q) = \frac{(4\pi)^2}{Q^4} \left| \int_{-\infty}^{\infty} \frac{d\beta}{dz} e^{iQz} \, dz \right|^2 .$$

In this limit, the loss of phase information results in:

(a) Translational invariance: $\beta(z)$ and $\beta(z+z_0)$ give identical data.

(b) Inversion invariance: $\beta(z)$, $\beta(-z)$ and $-\beta(z)$ give identical data.

(c) Symmetric solutions, similar to an autocorrelation or Patterson function, often fit the data well. (Note that linear combinations of solutions, $\beta(z)$ & $\beta(-z)$ for example, are not solutions in general.)

(d) Physically, and chemically, nonsense solutions sometimes fit the data.

(e) The program can easily get stuck in a local solution (c.f. Fig. 2).

Translational invariance is of little practical consequence and the inversion invariance can usually be dealt with using prior chemical and physical knowledge about the sample and its preparation. Although a simulated annealing algorithm (Pannetier, 1990) can be used to avoid the problem of getting stuck in local subsidiary solutions (at least in principle), it cannot help in the case when there are many almost equally probable "global" solutions (c.f. Fig. 2).

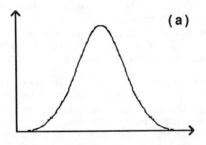 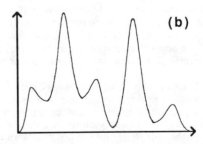

Figure 2: A schematic 1-d illustration of likelihood functions. The horizontal axis is a pseudo multi-dimensional axis representing various possibilities for the quantity of interest (for reflectivity, this would be various possible depth profiles); the vertical axis is the probability of obtaining the measured data-set given the quantity of interest (the "fit to the data" term). **(a)** A well-behaved likelihood function, having a single maximum, common for simple linear problems. **(b)** The complex topology of a badly-behaved likelihood function, showing many (local & global) maxima, which arises for non-linear problems such as reflectivity.

3. Model-fitting and free-form solutions

The standard approach for reflectivity data analysis is one of model-fitting and refinement, as exemplified in Jeff Penfold's talk in these proceedings. One assumes a functional form for the depth profile, and estimates the parameters defining this model using a least-squares fit to the data. This is fine: in the Bayesian context, one chooses the *hypothesis space* (or set of possible answers for β(z)) defined by the parameters of the functional model and assigns a uniform prior p.d.f. over these parameters; if the likelihood function is of the form $\exp(-\chi^2/2)$, then the maximum of the posterior p.d.f. is given by the least-squares solution.

The great advantage of this standard approach is that it usually works! There are, of course, several reasons why it works: **(i)** it is a small problem because β(z) is described by only a few parameters; **(ii)** a lot of prior knowledge is used in choosing the functional model, and this automatically rules out many of the nonsense solutions; **(iii)** if one has a good starting model then the analysis begins much closer to the desired solution, thereby reducing the risk getting stuck in a local solution.

The big disadvantage of the standard approach is the flip side of its advantages — where did the functional model come from? Perhaps we have good reason to believe our model because we prepared the sample and know how the layers of slime and sludge were laid down. On the other hand, some interesting chemistry could have taken place and invalidated our knowledgeable assumptions. Perhaps we have an oil-bearing sample from under the ocean, which we obviously did not prepare — how are we to choose a model for its density profile? Despite our rather extreme example, the problem of having to choose a model to refine is quite common. It is usually dealt with by a laborious, and immensely time-consuming, process of trial-and-error. Can we do any better?

If we do not have a (good) model for the density profile, then we must try to obtain a "free-form" solution for β(z). We might, for example, digitise the z-coordinate into a suitably large number of pixels and try to estimate the value of β (flux) in each pixel. This has the advantage that very little is assumed about the nature of the density profile, but it has the serious disadvantage that it is a much tougher (if not, generally, impossible) problem. The difficulty of having many, very different, often silly, solutions being allowed by the data for a phaseless problem was mentioned in the last section — this problem is, of course, exacerbated by the increased flexibility allowed by the large number of parameters now used to describe β(z).

The severity of the situation is evident from work on the closely related phaseless Fourier problem. Using the work of Bruck & Sodin (1979), who consider the factorisability of the Fourier transforms in different analytic forms, some (Feinup 1984 and Bates & Mnyama 1986, amongst others) are brave (and perhaps foolish) enough to state that there is usually a unique solution for problems in 2 & higher dimensions if the object of interest is positive and of finite support. It is agreed, however, that the 1-dimensional case, such as reflectivity, is the most difficult in the sense of non-uniqueness and numerous ambiguous solutions.

Be that as it may, a necessary condition for having a finite set of solutions for any under-determined problem is the use of a non-uniform prior. For the case of model-fitting, we usually wish to estimate a handful of parameters from about 100 data — a uniform prior p.d.f. is fine in this instance because the data, for example in the form of least-squares fit, impose a reasonable constraint on the possible solutions. For the free-form approach, we are usually trying to estimate more parameters (pixel fluxes) than we have data — the simple least-squares method will just not work because there are an infinite number of possible solutions. We need some additional constraint on β(z) in order to remove this infinite ambiguity. In the Bayesian context we need to use a non-uniform prior p.d.f. for β(z), so that there is a preferred solution even in the absence of any data — we need a regularising function. Entropy is an example of a regularising function, and we illustrate its use for analysing (simulated) neutron reflectivity data in the next section.

4. Examples

Despite all our warnings of the dangers and difficulties of analysing neutron reflectivity data with the free-form solution approach, we tried doing it with simulated data using MaxEnt. When using MaxEnt we assume a prior p.d.f. of the form prob[$\beta(z)$] $\propto$ exp(αS), where the generalised Shannon-Jaynes entropy S is given by (Skilling 1989):

$$ S = \sum_j \beta_j - m_j - \beta_j \log\left(\frac{\beta_j}{m_j}\right) \quad , $$

where β_j is the flux in j^{th} pixel of the digitised reconstruction of $\beta(z)$ and $\{m_j\}$ is a default model, or the solution to which the analysis will default in the absence of any data. The most naive choice for the default model, and the one we use for the examples in this section, is a featureless or flat default: $\{m_j\}$ = constant. Although the regularising parameter α should also be chosen by Bayesian methods (Skilling 1989 & 1990), we will use the historical recipe of choosing α such that the solution fits the data according to the "χ^2 = the number of data" criterion (Gull & Skilling 1984).

The simulations were carried out on a grid of 100 points for $\beta(z)$, with a resolution of 10 Å, and 77 data evenly distributed between $0.01\text{Å}^{-1} \leq Q \leq 0.2\text{Å}^{-1}$. A Gaussian resolution function with $\delta Q/Q$ = 5%, in terms of a full-width-half-maximum, was used and some random noise added to generate "typical" data-sets. We analysed these data using MaxEnt, with a (low) flat

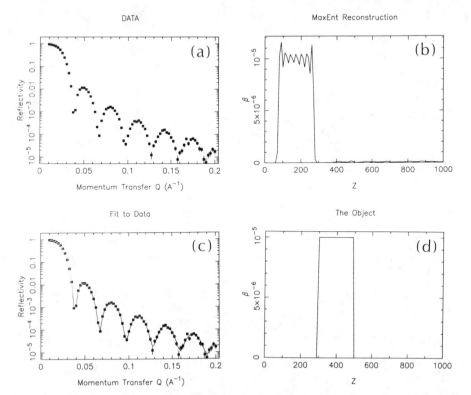

Figure 3: (a) Simulated reflectivity data. **(b)** The MaxEnt reconstruction of the depth profile. (z is in units of Å, and β is in units of Å^{-2}.) **(c)** The fit to the data. **(d)** The (true) test profile.

(uninformative) default model, and compared the answers with the test profiles. We were trying to be as dumb and naive as possible in order to see if we could get anywhere at all! We emphasise that this procedure is not recommended for the definitive answer to reflectivity data, but hope that it will be useful for providing a guide towards a suitable functional model to refine when no such model is available (or justified) a priori.

Fig. 3(a) shows the first data-set and Fig. 3(b) the resultant MaxEnt reconstruction of the density profile. Fig. 3(c) is the fit to the data and indicates that the MaxEnt solution is perfectly acceptable as far as the data are concerned. Fig. 3(d) was the true object, or test profile. We notice several features when comparing the true profile with the MaxEnt reconstruction. (i) There is a shift along the z-axis; this is not surprising since we cannot fix the origin when we do not have phase information. (ii) There is "ringing" in the reconstruction for large values of the scattering-length density ($\beta \sim 10^{-5}$). Ringing is a common consequence of having incomplete and noisy data (which is always the case in real life), and entropy reduces these artefacts largely through the power of positivity. The positivity constraint is very powerful when β is small ($\beta \sim 10^{-7}$, in this example), but it does not help much when β is large. A formal error analysis (Skilling 1990) would presumably indicate that these ringing features were not statistically significant, and the newer "pre-blur" ideas presented by John Skilling in these proceedings may help to reduce their visual impact. (iii) The most important observation is that the true structure has, essentially, been recovered. We could easily use the MaxEnt reconstruction as a basis to recommend refining a functional model consisting of a single slab of material; we would even have a very good starting estimate of the width and scattering-length density of this slab.

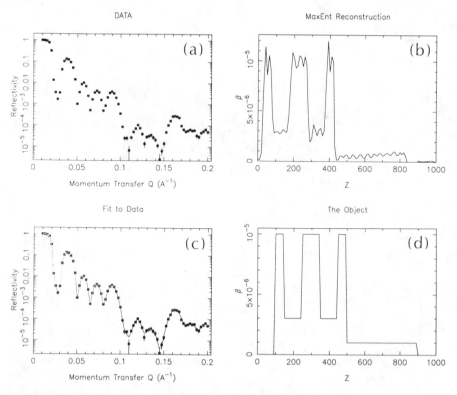

Figure 4: (a) Simulated reflectivity data. **(b)** The MaxEnt reconstruction of the depth profile. **(c)** The fit to the data. **(d)** The test profile.

Fig. 4 shows the results of using the same analysis on data generated from a more complicated test profile. The conclusions are much the same: there is z-axis shift, a certain amount of ringing artefacts, but the structure has again be recovered. Fig. 5 shows the results from another test profile. In addition to the previous features, we notice that the profile is back-to-front! As we noted in Section 2, at least in the Fourier limit, the data cannot distinguish between the correct orientation and its reverse mirror image. This ambiguity aside, again, the most important point is that the true profile has essentially been recovered.

Fig. 6 shows that all this was, of course, too good to last. The MaxEnt reconstruction of Fig. 6(b) is not a good guide to the true density profile, although we can see some resemblance of a symmetrised version of the test object. Notice, however, that the fit to the data is excellent! Note also that the test profile of Fig. 6 is much simpler than those of Figs. 4 & 5 — it is difficult to predict, a priori, to what level of complexity of $\beta(z)$ the method will work.

Rather than just giving up on this example, let us consider if can use any other piece of simple prior knowledge about $\beta(z)$ to constrain the reconstruction. Well, let us suppose we know the substrate and, hence, the value of $\beta(z)$ on the right-hand-side; suppose also that we know that the thickness of the sample on top of the substrate is less than 600Å. If we incorporate this information into our analysis, we obtain the MaxEnt reconstruction shown in Fig. 7(a). It is an improvement over the reconstruction of Fig 6(b), but is still some way off the truth. Let us continue further along this track: suppose we knew the thickness of the surface layers. This might well be possible if we also had x-ray reflectivity measurements of the same sample. For example, if it was a layered hydrocarbon sample in which successive layers had been deuterated to produce strong neutron contrast, the x-rays would usually see a roughly homogeneous slab since the electron density contrast between most hydorcarbons is small; the

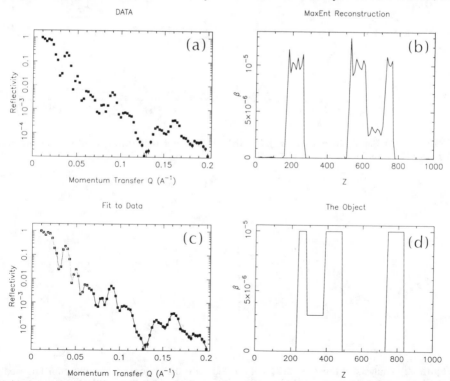

Figure 5: (a) Simulated reflectivity data. **(b)** The MaxEnt reconstruction of the depth profile. **(c)** The fit to the data. **(d)** The test profile.

data would look similar to the uniform fringes seen in Fig. 3(a), and so we could estimate the thickness of the slab. If we incorporate the knowledge of the substrate and the 400 Å thickness of the surface layers into the analysis, we obtain the MaxEnt reconstruction shown in Fig. 7(b). This still leaves something to be desired, but is sufficiently close to the truth to provide a good guide towards a suitable functional model.

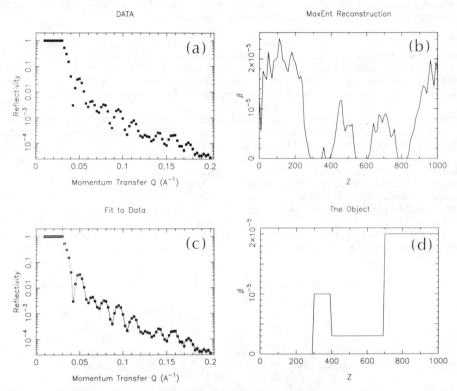

Figure 6: (a) Simulated reflectivity data. **(b)** The MaxEnt reconstruction of the depth profile. **(c)** The fit to the data. **(d)** The test profile.

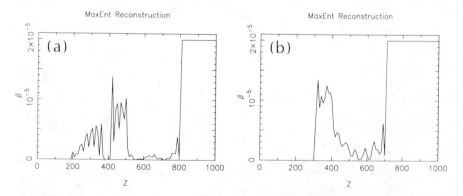

Figure 7: Analysis of the reflectivity data in Fig. 6(a) using some additional prior knowledge. **(a)** Given the substrate and that the sample is less than 600Å thick. **(b)** Given the substrate and that the sample is 400 Å thick.

5.   Conclusions

We have made a preliminary investigation of the potential use of MaxEnt to analyse neutron reflectivity data. The results are very encouraging. Our success rate was certainly far greater than we had feared considering the numerous pitfalls involved in obtaining a free-form solution for this difficult problem. In practice we should use as much (bone fide) prior knowledge about the sample as possible to help us, and have indicated how this might be done in the case when we have complementary x-ray reflectivity data.

In Section 3 we said that the use of a non-uniform prior p.d.f., in the form a regularising function for example, was necessary to obtain a free-form solution. Although we have used an entropic regularisation, it is not clear that this specific form is much better than simpler alternatives like Gaussian regularisation for the case of the density profile. This is because a lot of the additional power of entropy comes from the enforced positivity, a constraint which is often not very important for the density profile and sometimes not even true. Although negative scattering-length densities can be handled in the MaxEnt framework by using two-channel entropy (Laue et al. 1985), much of its useful power is sapped.

Finally, we conclude by reiterating that we are extremely encouraged by our preliminary results and believe there is great scope for integrating free-form solutions with model refinement for the analysis of neutron reflectivity data. However, we caution that, in general, there are no guarantees with the phaseless problem!

Acknowledgements

We thank Roger Pynn for his encouragement and useful discussions. This work was supported by the Office of Basic Sciences of the the U.S. Department of Energy.

References

Bates, R.H.T. & Mnyama, D. (1986). *Advances in Elec. & Electron. Phys.*, 67, 1-64.

Bruck, Yu.M. & Sodin, L.G. (1979). *Opt. Commun.*, 30, 304-308.

Feinup, J.R. (1984). *Indirect Imaging*, J.A. Roberts ed., Cambridge University Press.

Felcher, G.P. (1988). *Thin-Film Neutron Optical Devices*, SPIE, 983.

Gull, S.F. & Skilling, J. (1984). *IEE Proc.*, 131F, 646-659.

Jeffreys, H. (1939). *Theory of Probability*, Oxford University Press. Fourth edition: 1983.

Jaynes, E.T. (1986). *Maximum Entropy and Bayesian Methods in Applied Statistics*,
        J.H. Justice ed., Cambridge University Press.

Laue, E., Skilling, J. & Staunton, J. (1985). *J. Mag. Res.*, 63, 418-424.

Lekner, J. (1987). *Theory of the Reflection of Electromagnetic and Particle Waves*,
        Martinez Nijof Publishers, Dordrecht, Holland.

Pannetier, J. (1990). In these proceedings.

Penfold, J. (1990). In these proceedings.

Russell, T.P. (1990). Submitted to *Materials Research Reports*.

Skilling, J. (1989). *Maximum Entropy and Bayesian Methods: Cambridge 1988*,
        J. Skilling ed., Kluwer Academic Publishers.

Skilling, J. (1990). In these proceedings.

Inst. Phys. Conf. Ser. No 107: Chapter 5
Paper presented at Neutron Scatt. Data Anal. Conference, Rutherford Appleton, 1990

233

Can we justify conventional SANS data analysis?

R. E. Ghosh and A. R. Rennie

Institut Laue Langevin, 156X, F-38042 Grenoble Cedex, France

ABSTRACT: We have added time-of-flight analysis to D17, a long wavelength monochromatic beam SANS spectrometer. Performing small-angle measurements with this facility on a number of classical samples has allowed us to decompose our data into elastic and inelastic components. While justifying the traditional methods of analysis, these data clearly show features which are ascribed to inelastic scattering processes occurring in samples, or secondary standards. In addition some unexpected aspects of detector response are seen. Similar effects arise in experiments performed on pulsed source instruments.

1. INTRODUCTION

The success of small angle scattering as a technique in such diverse fields as materials science and biology is shown here to depend on a fortunate lack of discrimination in conventional instrumentation. We are able to demonstrate as a consequence of introducing energy analysis into the measurement process that the major part of the background signals measured often correspond to neutrons which have suffered considerable thermalisation. Use of transmission measurements for correcting small angle scattering data has always been of fundamental importance. These measurements are now seen, to a first approximation, to account for both inelastic and elastic scattering, permitting subtraction to yield the required elastic coherent scattering cross-section.

It has been recognised that the scattering from water as a secondary calibrant in SANS experiments is complicated, and empirical functions for the wavelength dependence of the total cross-section have been developed by Jacrot (1976) and Ragnetti *et al* (1985). These are assumed to have no momentum transfer dependence. Only a few authors have remarked on the variation of the sample total cross-section with temperature. As pointed out by Maconnachie (1984) this may have a significant effect in the data corrections.

The experiments described here indicate necessary precautions to be taken in interpreting conventional results, and especially the need to appreciate the details of the dynamics of samples and possible standard calibrants. First the instrument set-up is described. The time-of-flight method used for energy analysis shows up, in addition, a number of characteristic features of the detector system. The set of measurements corresponding to a standard small-angle scattering experiment is then presented. The data analysis is performed both with specific regard to elastically scattered neutrons and conventionally (i.e. ignoring any energy change of the neutron). The comparison of the results shows small but significant deviations.The individual spectrum components as a function of time of flight are then examined. The discussion draws on these spectra to explain the potential sensitivity of conventional small angle scattering results to the details of the internal motions of the samples under study.

2. EXPERIMENTAL

Measurements were made on the D17 small-angle neutron spectrometer at the high flux reactor of the I.L.L. in Grenoble (Blank and Maier, 1988). This instrument is shown schematically in figure 1. A monochromatic (10%) beam of wavelength 12 Å was obtained from the H17 neutron guide after passing through the helical slot velocity selector. The beam is defined by diaphragms before the sample. The planar two-dimensional detector (128x128 cells each $0.5x0.5cm^2$) was placed 1.40m from the sample, and rotated 6° horizontally about the sample to increase the angular range accessible in a single measurement.

For this experiment an additional disk chopper was mounted on the table 0.25m before the sample. To reduce the data to more manageable volumes the detector was encoded as 16x16 cells of $4x4cm^2$ using 64 time channels. Time-of-flight electronics triggered from a magnetic pickup on the chopper were set to record a full spectrum with a channel width of 94μsecs. The duty cycle of the pulsed monochromatic beam was 2.5%.

Elastic scattering is measured in between time channels 41 and 51. We compare this sum to the total to quantify the fraction of scattering occuring with no change in neutron velocity. The fractions for water 1mm, 25C, and toluene are 53% and 48% respectively. The standard D17 detector has an active depth of 1cm of $^{10}BF_3$ gas at 1.1 bar pressure. The detector efficiency was calculated as a function of energy based on these parameters, giving approximate values of 11% at 25meV, and 53% at 0.56meV (12Å). There is therefore some built-in discrimination in normal long wavelength experiments against inelastic processes where the incident slow neutrons gain energy. We have introduced these corrections in the summary table to estimate the true fraction of elastic scattering which actually occurs (independent of specific detector characteristics). Though important, the fine detail of this correction is not fundamental to the commentary presented here.

We note that placing the detector off-axis and then regrouping data in annuli about the beam axis averages the detector sensitivity which is known to be uniform about the geometric detector centre, but increases slightly as a function of radius from that point. The data were regrouped in a number of fashions to investigate this effect. The longer flight path to the edge of the planar detector was evident in the arrival time of the elastically scattered neutrons.

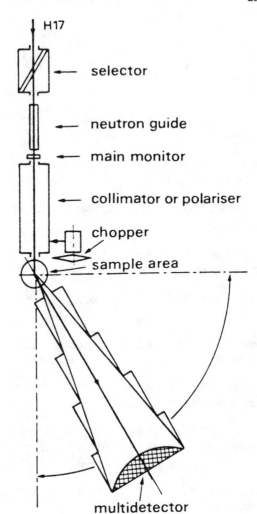

H17

selector

neutron guide

main monitor

collimator or polariser

chopper

sample area

multidetector

Fig. 1: Schematic diagram
of the D17 Spectrometer.

Several samples, summarised in table 1, were measured to provide data typical of small angle scattering experiments. Water, frequently used as a calibrant was measured at two thicknesses and two different temperatures. Figure 2 shows the raw data for 1mm of water, illustrating the two-dimensional spatial distribution for each time channel. The time scale for the arrival of the neutrons starts at zero in the bottom left-hand corner, and increases in rows from left to right. The most intense scattering corresponds to elastic scattering of the 12Å incident neutrons (about time channel 46). The corners of the detector are not neutron sensitive. The detector is 6 degrees off-axis and hence the beam-stop, which protects the detector from the direct beam shows up as white spot on each of the 64 time frames. A significant fraction of intensity is clearly evident at flight times other than those corresponding to elastic scattering.

Table 1.

Sample	Angle	SUM	SUM corrected	%Elastic corrected	Trans	Tm Neutron Gas temp K
Water 1mm 298K +cell+windows	8.5	54863	105770	28	0.418	436
Water 2mm 298K +cell+windows	8.5	66246	153700	17	0.178	446
Water 1mm 338K +cell+windows	8.5	53392	112060	23	0.375	457
Water 2mm 338K +cell+windows	8.5	61074	157650	13	0.143	481
Empty Cell 298K +windows	8.5	5958	9530	29	0.975	211
Vanadium .65mm +windows	8.5	7168	10260	42	0.858	-
Toluene 298K +cell+windows	8.5	46173	86370	26	0.503	252
5% PS-d in Toluene +cell+windows	2.5	94711	136780	51	0.510	
	8.5	63289	101830	39	0.510	254
	14.5	57486	98250	32	0.510	
Empty Hole +windows	8.5	5420	8460	29	1.000	

Note: SUM is the sum of counts per hour per $140cm^2$ detector surface (8.5 ± 1°)
The corrections correspond to an energy dependent detector efficiency

    While rarely used on D17 due to the high absorption cross-section, and unreproducibility of specimens, a sample of vanadium was measured to give a reference spectrum of a predominantly isotropic elastic scatterer, having a Debye temperature of 320K. Appropriate empty cells were included in the set. The sample measured for coherent small angle scattering was a 5% solution of perdeuterated polystyrene (PS-d) in hydrogenous toluene, and an equivalent toluene background. The sample of polystyrene was purchased from the Polymer Standards Service, Mainz, and characterised by g.p.c. giving M_W = 13600, M_N = 12800.

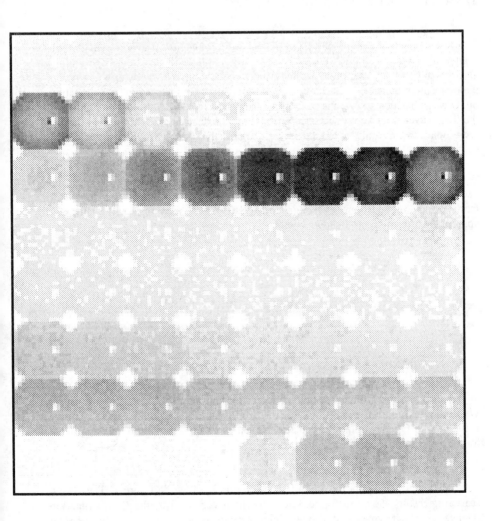

Fig. 2: A set of 64 two-dimensional spectra for 1mm water (25 C) as a function of time-of-flight (raw data). The grey scale goes from light to dark with increasing intensity. The time scale for neutron arrival starts at short times in the bottom left hand corner and increases by rows from left to right and from bottom to top. The most intense signal corresponds to elastic scattering of 12 Å neutrons (time channel 46). The detector is 6 degrees of axis and the beam stop protecting the detector from the incident beam is off-centre showing as a white spot on each frame. The large fraction of inelastically scattered neutrons is evident from the intensity at short times of flight.

3. SMALL ANGLE SCATTERING RESULTS

The data were regrouped for analysis in the conventional way for small-angle scattering. Two procedures were used: first only the time channels corresponding to elastic scattering were selected, and the data summed to give intensity as a function of the momentum transfer, Q. Secondly all time channels were included in the summation, but Q was still calculated on the basis of an unchanged incident wavelength, rather than the final wavelength which was known from the time distribution. This second method corresponds to the treatment of total scattering normally made in SANS experiments.

The sample data S(Q) were corrected in the usual way (Ghosh 1989) for background scattering B(Q), and normalised to the scattering from 1mm water W(Q), from which an appropriate background WB(Q) had been removed, using the appropriate transmissions T shown in table 1:

$$S_{cor}(Q) \ = \ \frac{S(Q) - \frac{T_S}{T_B} B(Q)}{W(Q) - \frac{T_W}{T_{WB}} WB(Q)} \cdot \frac{T_W}{T_S}$$

to give the scattering from the sample corrected for detector efficiency, and solid angle. We note again that offsetting the detector from the beam axis reduces the effective variation of efficiency of the flat detector. Nonetheless some increase is still significant for the water samples at larger Q values.

The corrected data for the polymer solution is shown in figure 3. The elastic scattering results shown in figure 3a can be fitted very well by the Debye function (Debye 1947), shown as a solid line, giving a radius of gyration of 24.7Å. The total scattering data is shown in figure 3b. The best fit (solid line) departs significantly from the data at larger Q, corresponding to the overcorrection due to increased total scattering from the water used in the normalisation procedure. The corresponding value of the radius of gyration was 25.9Å. For comparison here, the broken curve shows the correct model derived from the elastic scattering measurements after being rescaled. The total scattering data has a distinctly different shape from that obtained in the elastic anlysis only.

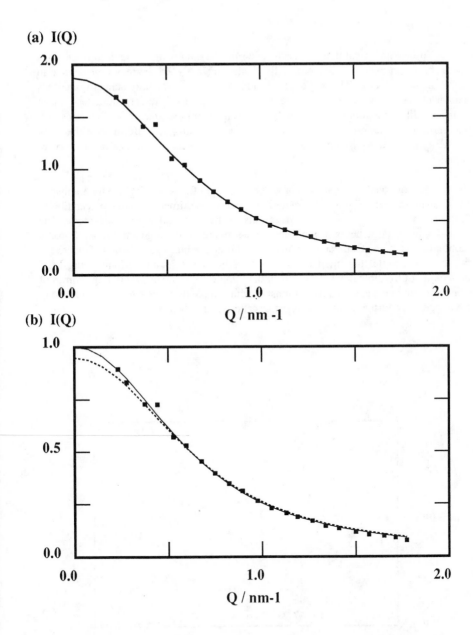

Fig. 3: (a) Elastic scattering for PS-d in toluene with fit of Debye function and (b) total scattering, all time channels, with best fit to data (solid line) and the model fitted to elastic data rescaled (dashes) The latter data show the deviation at large Q values due to an over correction with water.

4. TIME-OF-FLIGHT RESULTS

To facilitate treatment for each spectrum the data were grouped into ten rings about the beam axis. The detector cells in the ring 7.5 - 9.5 degrees represented an active area of 140cm² These grouped spectra were then treated by standard methods similar to that above, using the same transmission factors. The vanadium served as a primary calibrant exhibiting principally elastic incoherent scattering at 3000μsecs m⁻¹, and permits calibration and normalisation of the remaining data. Because of the weak scattering it was used in this case as a cross-check on elastic peak data from the other samples.

The corrected data for 1mm of water are seen in figure 4 to be heavily dominated by inelastic scattering with a characteristic peak in the time-of-flight distribution at 360μsecs m⁻¹ (40meV), which corresponds to the torsional vibrational mode. The cross-section is large being proportional to the square of the large amplitude of proton motion (Zemach and Glauber 1956 and Reynolds and White 1969). When compared to the thicker sample this intensity scales directly with thickness (there being little probability of multiple inelastic events in the sample). The intensity at longer times of flight does not have any simple relationship with sample thickness. As expected raising the temperature further reduces the elastic scattering at 3000μsecs m⁻¹.

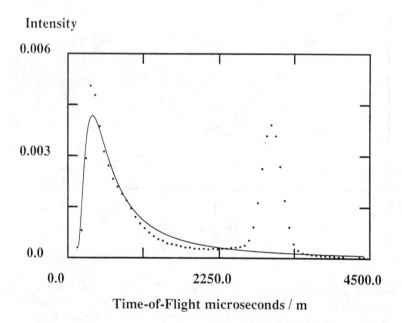

Fig. 4: Time-of-flight spectrum of 1mm water (25 C) at 4.5 degrees with fit of simple Maxwellian gas model with a characteristic temperature Tm = 436 K.

The background due to the fused silica (used for windows to the evacuated detector tube and the sample cell) is dominated by inelastic "phonon" scattering. The incident wavelength is sufficiently long that few elastic coherent processes can contribute. However, since these windows are at room temperature, vibrational de-excitation can occur transferring appreciable energy and momentum to the incident neutron. This cross-section is small compared to the samples.

Toluene (points in figure 5) shows a smoother spectrum than water. There are no large amplitude molecular modes at low energies apart from the methyl rotational modes which form a continuum, and to which the inelastic scattering corresponds.

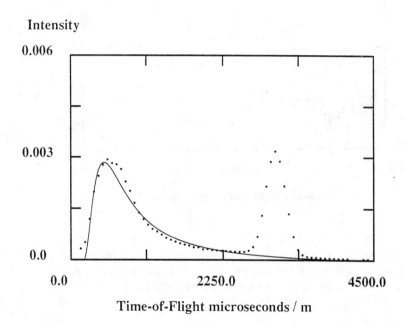

Intensity

Time-of-Flight microseconds / m

Fig. 5: Spectrum of 1mm toluene at 4.5 degrees, this is distinctly smoother than water, the line corresponds to a Maxwellian curve with Tm = 252 K.

5. DISCUSSION

Figure 6 shows in an exemplary fashion that the standard subtraction (dashed line) of the toluene background from that of the polymer solution (points) leads to very nearly complete elimination of the inelastic incoherent scattering yielding the remaining coherent elastic small-angle signal. The residual inelastic scattering (shown magnified) is structured, demonstrating that there is a slight difference in the motions of the solvent and solution. The transmission corrections should take into account the slight differences in cross-section due to the displacement of solvent by the deuterated solute. Transmission measurements only show the net loss from the incident beam, ignoring the fate of the scattered neutrons. Given the predominant inelastic scattering which occurs in conventional experiments there is vindication for carefully measuring transmisions at each temperature that sample data are taken to ensure the correct (inelastic) background is subsequently subtracted.

Intensity

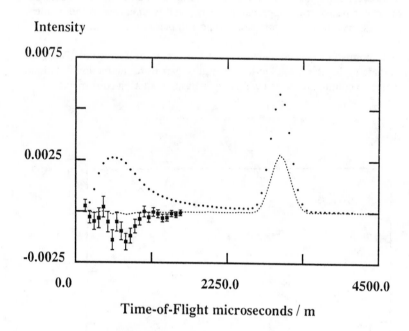

Time-of-Flight microseconds / m

Fig. 6: The spectrum of PS-d in toluene is shown as large points After conventional correction the dotted curve is obtained. The inelastic region is shown multiplied by 10 (with error bars) to show the residual mismatch in the resultant spectrum.

When using inelastic neutron scattering to elicit details of molecular dynamics every effort is usually made to use thin samples and special geometries to restrict the possible scattering to one event within the sample. With the SANS samples, especially the hydrogenous samples, the cross-section corresponds to a complex mixture of multiple processes, indicated by the small fraction of the total beam transmitted. Rather than attempt to model the samples it is simpler to consider the neutron, with an incident energy of 0.56meV, corresponding to a temperature of 7K, being *thermalised* as it penetrates the sample (or windows etc.). A single parameter Maxwellian function may then be used to model the resultant inelastically scattered neutrons, namely that the energy distribution N(E) corresponds to that of a simple gas at temperature Tm (Windsor 1981).

$$N(E) = \frac{E}{Tm^2} e^{-\frac{E}{Tm}}$$

It is found that this function, when suitably scaled approximates the observed inelastic distributions for all the samples and windows, and the temperatures derived for this neutron gas are listed in table 1. We show in figures 5 and 6 this function (solid lines) superimposed on the inelastic scattering region. As expected, for water the effect of sample thickness is seen to increase the fraction of neutrons thermalised, and the final temperature Tm of the neutrons rises as the sample is warmed. The characteristic temperatures from the water measurements are exceptionally high due to the intense torsion mode. This involves additional effects when measured with the standard detector. This present simplistic treatment is not sufficiently complete to explain quantitatively the correction functions formulated for water calibrants. This will be described in a further paper.

6. CONCLUSIONS

We note the advantages of performing regrouping by software, following the experiment. This enabled us to identify a number of artefacts which may be ascribed to the planar geometry of the detector. This has additional implications in experiments performed at high resolution, and with even longer wavelengths for the incident neutrons.

The effect of thermalisation has been shown to be crudely represented by a Maxwellian function for most materials. The derived characteristic temperature is sensitive to the nature of the sample and, of course the thickness. Solid samples require careful preparation of both samples and blanks with very similar thicknesses, as well as similar internal dynamical properties to enable the inelastic scattering to be subtracted. In the case of aqueous solutions the vibrations of the solute nuclei must also be taken into account. In general this requires measurement of an additional calibrant.

In the first approximation the total scattering cross-section assessed from the transmission measurements will allow background data to be correctly subtracted. The effect of the thermalisation, to which a considerable fraction of neutrons are subject, should not be ignored since neutron thermal equilibration is never identical or complete. This is of particular pertinence to the water as a calibrant, where it is clear that the normalisation of the energy sensitive detector is made conventionally with a wide distribution of incident neutron energies, though the corrections sought are for the long wavelength elastic coherent component only. While leading to erroneous values of only a few percent for values of I(0), and radius of gyration in the favourable case above (comparable in magnitude to ignoring finite instrumental resolution), greater caution is required in the interpretation of models weighted at higher Q, e.g. surface area measurements (Porod's Law), or in invoking Fourier transformations (q.v. Glatter 1982). Additional problems may arise in interpreting data obtained through sequences of measurements using different degrees of deuteration.

While this effect is relevant to measurements performed on long wavelength instruments, where, as we have shown, the elastic scattering can be accurately measured by using a chopper to discriminate against neutron energy gain, similar energy changes occur on small angle instruments using shorter wavelengths. When instruments employ white incident beams (Windsor 1981b) thermalisation in the sample also occurs. In these cases most information is obtained at shorter wavelengths, often less than 4Å (1000μsecs m^{-1}). With a fraction of resultant neutron velocity distributions after the samples similar to that observed here for thermalisation processes it is clear that inelastic scattering will lead to a distortion in the regrouping of data assigned to specific wavelengths.

ACKNOWLEDGEMENTS

We wish to thank J. Foubert for his modification of the CERN program PAW used to read, treat, and examine the ILL data in 256 colours, and to M. Cruz for his help in mounting the chopper facility used in this experiment.

REFERENCES

Blank H. and Maier B. (eds) 1988 *Guide to Neutron Research Facilities at the ILL*
  ILL Grenoble
Debye P. 1947 *J.Phys. Colloid. Chem.* **51** 18
Ghosh R. E. 1989 *I.L.L. Internal Report* 89GH02T
Glatter O. 1982 *Small Angle X-ray Scattering,* ed. O. Glatter and O. Kratky,
  (London: Academic Press) Chapters 4,5
Jacrot B 1976 *Rep. Prog. Phys.* **39** 911
Maconnachie A. 1984 *Polymer* **25** 1068
Ragnetti M., Geiser D., Höcker H. and Oberthür R.C. 1985 *Makromol. Chem.* **186**
  1701
Reynolds P.A. and White J.W. 1969 *Discuss. Farad. Soc.* **48** 131
Windsor C.G. 1981a *Pulsed Neutron Scattering*, Taylor and Francis, London p113
Windsor C.G.1981b loc. cit. p258
Zemach A.C. and Glauber R.J. 1956 *Phys. Rev.* **101** 118

Inst. Phys. Conf. Ser. No 107: Chapter 6
Paper presented at Neutron Scatt. Data Anal. Conference, Rutherford Appleton, 1990

245

CLIMAX: A program for force constant refinement from inelastic neutron spectra

G. J. Kearley
J. Tomkinson

Institut Laue Langevin, B.P. 156X, 38042 Grenoble, Cedex, France.
Rutherford Appleton Laboratory, Chilton, Didcot, Oxon OX11 0QX

ABSTRACT
CLIMAX provides a method by which a vibrational analysis of molecular modes can be performed by the refinement of force constants to give a best least squares fit of a calculated INS spectrum to the observed INS spectrum. This idea has been extended to operate using either cartesian, internal or symmetry coordinate systems and to include the effects of recoil in either "stiff" or "weak" lattices. Visualisation of the individual atomic vibrational displacements for a given mode using "ORTEP" is now also provided.

1. INTRODUCTION

CLIMAX is a program which attempts to reproduce the observed inelastic neutron scattering (INS) spectrum of molecular vibrations in wide variety of species. The original version was written in 1985[1] and showed that it was feasible to refine interatomic force constants directly on the basis of the observed INS spectral profile. Since then treatment for inverted time-of-flight spectrometers has been introduced[2], overtones and combinations have been introduced[3], and the effects of recoil accommodated[4].

During this period of development CLIMAX has been exposed to a wide variety of users and several analyses have been achieved and published directly from its use. It has become clear that CLIMAX has to cater for a wide variety of users and most of this paper is devoted to describing how we have tried to meet this need. In the most basic case the user wishes only to know if the observed INS spectrum is consistent with a proposed molecular model, and in all probability, the user will be neither a specialist in vibrational spectroscopy nor neutron scattering. At the opposite extreme specialists in vibrational spectroscopy wish to have access to all the rigors of the normal coordinate analysis. Further, for molecular species of low mass the effects of recoil have to be considered, particularly by neutron scatterers who are pushing more towards long-range effects.

In order to cope with this diversity, we have created several options. It is easiest to consider first the various routes by which the vibrational frequencies and the mean square displacements can be obtained and then secondly show how the calculated INS spectrum can be obtained from this information for different types of system.

2. VIBRATIONAL ANALYSIS OPTIONS

There are three coordinate systems which can be used for the vibrational analysis in a combined or an individual way.

2.1 Cartesian Coordinates.

Solving the vibrational problem in cartesian coordinates insulates the user from the use of group theory and is thus well suited to non-specialist users, particularly for complex molecules. This route through the normal coordinate analysis is shown schematically in Figure 1 by the bold arrows. Although this route may look shorter the matrices are much bigger since there are many redundancies. Further, as there is no symmetry information there is no block diagonalization of the matrices and it is impossible to make any symmetry constraints in the fitting procedure.

2.2 Internal Coordinates.

The original version of CLIMAX solved the vibrational secular equation in internal coordinates because it was thought that this coordinate system, which uses bonds, bond angles etc. as its definitions, was conceptually simple. To allow block diagonalization of the various matrices and to enable observed frequencies to be matched directly with calculated modes of particular symmetries, symmetry coordinates were also made available. This route is shown by the simple arrows on Figure

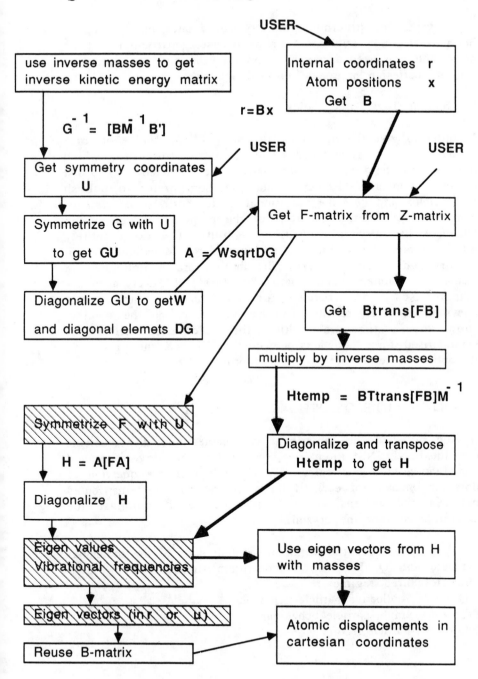

Fig. 1. Scematic illustration of the optional coordinate systems of CLIMAX.

1. However, constructing symmetry coordinates has proved to be a frustrating experience for many users, particularly if no constraints are needed anyway, and there is now the option to use the cartesian approach outlined above.

2.3 Symmetrised Force Fields.

In order to give the maximum flexibility in the choice of force field, CLIMAX always uses the Z-matrix formalism which defines how the force-constant matrix is constructed from a list of force constants. However, many authors prefer to publish force fields and describe vibrational modes entirely in terms of symmetry coordinates. Clearly, for the purposes of comparison CLIMAX had also to offer the possibility of working entirely with symmetry coordinates, so that force constants, eigen vectors and potential energy distribution are all obtained in terms of symmetry coordinates. In fact, CLIMAX normally symmetrises the force constant matrix and then unsymmetrises the results. The hatched boxes on the diagram show those stages at which the calculation differs if symmetrised force constants are input. CLIMAX then gives all its output in terms of symmetry coordinates.

3. MOLECULAR RECOIL OPTIONS

Having obtained the vibrational frequencies and displacements the idealized INS intensities of fundamental, overtone and combination modes is calculated. There is then the question of how the intensity of each of these "peaks" will be redistributed between the internal modes and the recoil amongst the external modes of the crystal.

We may write P as the initial momentum state of the molecular partticle and Q the neutron momentum transfer. Then theory identifies three regimes:

$P/Q \gg 1$; collective particle behaviour

$P/Q \sim 1$; single particle behaviour

$P/Q \ll 1$; Particle recoil

3.1 P/Q >>1

Here the molecular species are massive or, more rarely, extensively bound to each other. Thus, recoil can be neglected, and the final calculated spectrum is generated by convoluting the spectrum of Gaussian internal-mode peaks with the instrumental response function[5].

3.2 P/Q ~ 1

This intermediate case is the most difficult to treat in that the calculated internal-mode intensity is shared between the "phonon wings" and the band origin[4]. Unlike the other cases, these phonon wings are not a structureless hump but are the a multiphonon external mode spectrum which has structure. The external mode spectrum may be difficult to measure if it is overlaps the low-frequency region of the internal modes. Fortunately, we can normally isolate the one-phonon external mode spectrum. From this we generate a reasonable approximation to the full multiphonon spectrum. This is achieved in the harmonic approximation. The shape of the nth multiphonon contribution is the nth auto-correlation of the one-phonon spectrum. The simplicity of this method and its remarkable success (at least for simple systems) makes this an excellent first approximation. Only a single parameter (the external Debye-Waller factor) need be refined to determine the phonon-wing intensities.

3.3 P/Q << 1

In this case the intensity arising from an internal mode originating at ν is recoiled to $(\nu + E_r)$. The extent of recoil is related to the particles Sachs-Teller mass, μ. Then:

$$\nu/E_R = \mu - 1$$

The width of this Gaussian is related to this shift and to the kinetic energy, both of which are the same for all bands and can be refined.

At this point we stress that the mean square displacement $<u^2>$ of the scattering atoms are now completely determined. The values obtained from CLIMAX can be compared with diffraction results (after due allowance for any temperature differences). If any significant differences arise between estimated values of $<u^2>$ provided by the two techniques, these will be a direct result of differences between static and dynamic disorder.

4. VISUALIZATION OF MODES

There are several ways in which the calculated vibrational modes can be presented. Written descriptions are usually in terms of symmetry coordinates, although internal coordinates can also be encountered. CLIMAX provides both of these but it is not always simple to visualize what the vibration actually looks like. Graphical descriptions use displacement vectors but even these can be confusing for complicated systems and it is difficult to relate these if the "view point" of the presentation is changed. In an attempt to overcome this CLIMAX outputs an "ORTEP" file for each vibrational mode. This file contains the static positions of the atoms and their displacements transformed into anisotropic temperature factors. ORTEP[6] (or other packages) can then be used to plot and manipulate the molecule with the vibrational ellipsoids appearing in place of the thermal ellipsoids. The advantages of using existing crystallographic packages are obvious. An example of the three modes of the bridging hydrogen atom in the $H_5O_2^+$ ion are shown in the Figure in which the appearance of the vibrational modes is immediately apparent.

A further advantage comes from this approach. Not only can each mode be visualized from any "view point", but also the sum of the displacements for all modes - including the isotropic external mode contribution if phonon wings were present. This allows a further constraint on the vibrational analysis if the crystal structure is known since although we would not expect the total vibrational and thermal ellipsoids to superimpose we would at least expect that their major axes would coincide[7].

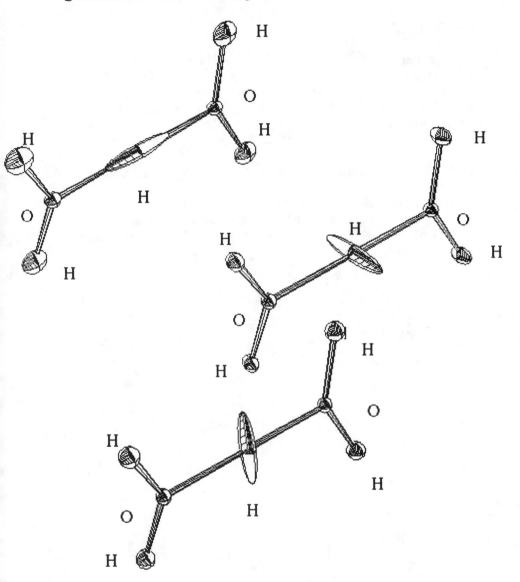

Fig. 2. Atomic displacements during the molecular vibrations of the hydrogen-bond bridge in the H5O2+ ion.

5. REFERENCES

1. G. J. Kearley. *J. Chem. Soc., Faraday Trans. 2* **82**, 41 (1986).
2. G. J. Kearley and J. Tomkinson. *Vibrational analysis of fundamental, overtone and combination molecular modes by profile refinement of INS spectra from spectrometers* (Adam Hilger, Bristol, 1986).
3. G. J. Kearley, J. Tomkinson and J. Penfold. *Z. Phys. B* **69**, 63 (1987).
4. J. Tomkinson and G. J. Kearley. *J. Chem. Phys.* **91**, 5164 (1989).
5. H. J. Lauter and H. Jobic. *Chem. Phys. Lett* **108**, 393 (1984).
6. C. K. Johnson. *ORTEP* (Oak Ridge National Laboratory, 1970).
7. G. J. Kearley, R. P. White, C. Forano and R. C. T. Slade. *Acta Cryst.* **In Press**, (1990).

Inst. Phys. Conf. Ser. No 107: Chapter 6
Paper presented at Neutron Scatt. Data Anal. Conference, Rutherford Appleton, 1990

253

Set-up and optimization of scans with the rotating analyser crystal spectrometer ROTAX

H. Tietze[1,2], W. Schmidt[1], R. Geick[1], H. Samulowitz[3] and U. Steigenberger[4]

1 Physikalisches Institut d. Univ. Würzburg, Am Hubland, D-8700 Würzburg, F.R.G.
2 Institut f. Festkörperforschung der KFA Jülich, Pf 1913, D-5170 Jülich, F.R.G.
3 Hewatt GmbH, Am Hahnenkreuz 46, D-5190 Stolberg-Dorff, F.R.G.
4 Rutherford Appleton Laboratory, Neutron Division, Chilton, Didcot, OX11 0QX, U.K.

The ROTAX spectrometer is an inverted geometry crystal analyser time-of-flight spectrometer for coherent neutron inelastic scattering on single crystals. It will be installed at the N2 beam-line of ISIS. Recently, first neutron scattering tests on a test assembly of ROTAX have been performed at ISIS giving very promising results about the principal scan performance and flexibility of ROTAX. However, mathematical refinements on the initially desired scan functions are required for optimising the accuracy of the regulation government. In this paper we discuss the influence of these refinement procedures on the actual scan parameters and performance and present the first neutron scattering results obtained on ROTAX.

1. INTRODUCTION

ROTAX stands for rotating analyser crystal spectrometer and it will be a highly flexible time-of-flight neutron spectrometer of inverted geometry with one single crystal analyser. By means of a user defined, programmable non-uniform rotation of the analyser crystal several types of scans in (Q,ω)-space can be performed (Geick and Tietze 1986), e.g. const-energy or const-$\bar{Q}/Q$ scans. The scope of the instrument will be neutron inelastic scattering investigations on single crystalline samples. In some sense ROTAX will compete with triple axis spectrometers at reactor thermal beam lines and of course with conventional TOF-crystal analyser spectrometers installed at pulsed spallation sources, e.g. with the Constant-Q spectrometer at LANSCE (Robinson et al 1985) and with the high-symmetry spectrometers MAX at KENS (Tajima et al 1982) and PRISMA at ISIS (Andreani et al 1987, Steigenberger et al 1990). As discussed earlier (Tietze and Geick 1987) the major advantage of ROTAX in comparison to the above mentioned spectrometers is to provide the user with an instrument with a much higher degree of flexibility in the choice of scans.

In this paper we shall briefly introduce the basic ideas of the instrument and its technical design. The mathematical conditions and the basic logic units of the electronics will be described. We shall explain mathematical optimisation procedures to be applied on the actually performed experimental scans in order to gain stability and perfection. We consider our contribution to this Conference on Neutron Scattering Data Analysis, as an example of basic mathematical and computational aid to the set-up and technical performance of an instrument. This is admittedly a little bit off the general scope of the conference, where methods to better understanding and refinement of experimentally obtained neutron scattering data have been in the centre of interest. However, we are not yet heading for improved analysis of scattering laws of a sample as the unknown input function to a system like a neutron spectrometer. At the present stage of development we are rather dealing with well known input functions, which determine the scan perfomance of the ROTAX instrument, and we measure the response of the mechanics and electronics as a whole. This function is used to deconvolute the unknown response function of the system, then a modified input function is to be fed back in order to optimise the desired technical performance of the machine. The circumstances are somewhat different from the general purpose of an accurate and straight foreward analysis and interpretation of neutron scattering data, for which techniques such as maximum entropy methods (Skilling 1990) have turned out to be very useful and powerful. We rather have to use Fourier transform and linear response theory to provide a reliable machine as the first step, previous to neutron data interpretation.

2. THE PREMISE FOR ROTAX

The basic idea of the instrument is to analyse by Bragg reflection all the neutron energies being scattered during one ISIS pulse from a sample placed in a white pulsed beam. During the neutron flight time between the moderator and the sample, the neutron pulses will disperse in energy and time (cf. fig.1). According to the scattering function of the sample, the neutrons will then be scattered within a certain solid scattering angle ϕ into polychromatic sequences of neutron packages, which will reach the analyser crystal consecutively after a well defined time-of-flight. If the analyser crystal is then positioned at the appropriate time to its Bragg angle $\Theta_A(t)$ corresponding to the scattered neutron energy, it will scatter this particular neutron package into a wide angle position sensitive detector. The scattering equations involved have been determined earlier (Geick and Tietze 1986); the scattering angles ϕ and $\Theta_A(t)$ and the total time-of-flight determine the energy and momentum transfer of the neutrons unambiguously. In order to detect several (ideally all) of the scattered neutron energies, it is essential to turn the analyser crystal in an accelerated, non-uniform spinning mode which must be synchronized with the time structure of the neutron pulses. There is not one general function of analyser motion, but it rather varies from one scan through (Q,ω)-space to another depending on the initial scan parameters as being defined by the experimentalist.

Fig. 1. Schematic diagram
of ROTAX and neutron
the pulse propagation.
M: moderator, S: sample,
A: analyser, PSD: detector.

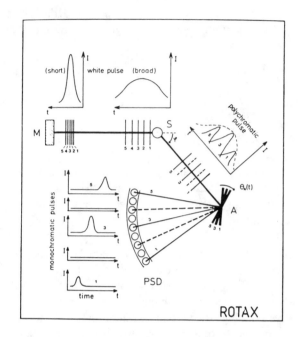

3. TECHNICAL CONCEPT

The non-uniform rotation function $\Theta_A(t)$ of the analyser crystal is of course periodic in time and must match the neutron source pulse repetition frequency of 50 Hz. Out of the total time interval of 20 msec for every neutron frame only a certain fraction will be specified by the scattering equations (Geick and Tietze 1986), i.e. when the neutrons with the desired incident energies are actually penetrating through the instrument. We shall denote this time interval the physical scan part. Within this the analyser Bragg condition is defined by the secondary flight time function $t_a(t_i)$, i.e. the scan dependent neutron time-of-flight from the sample to the analyser and further to the detector:

$$\Theta_A(t_a) = \sin^{-1}[\, K \, t_a(t_i)\,] \qquad \text{with} \quad K = h \, / \, (\, 2\, m_n \, L_a \, d_A \,) \qquad (1)$$

t_i is the initial time-of-flight from the moderator to the sample, h is Planck's constant, m_n the neutron mass, L_a the sample to analyser distance, d_A is the analyser's lattice spacing. More explicitly $t_a(t_i)$ can be written:

for const-energy scan:
$$t_a(t_i, \hbar\omega) = \frac{L_a}{L_i} \, t_i \left[\, 1 - \frac{2\, \hbar\omega \, t_i^2}{m_n \, L_i^2}\,\right]^{-1/2} \qquad (2)$$

for const-$\vec{Q}$/Q or
const-Ψ scans:
$$t_a(t_i, \Psi) = \frac{L_a}{L_i} \, \frac{\sin(\phi + \Psi)}{\sin \Psi} \qquad (3)$$

Ψ is the sample orientation angle, i.e. the angle enclosed by $\vec{k}_i$ and $\vec{Q}$. The analyser rotation dynamics are governed by the Bragg condition and its derivatives

$$\Theta_A(t_A), \quad \omega_A(t_A) = \frac{d\,\Theta_A(t_A)}{d\,t_A} \quad \text{and} \quad \alpha_A(t_A) = \frac{d^2\,\Theta_A(t_A)}{d\,t_A^2} \tag{4}$$

at the time $t_A = t_a + t_i$ when neutrons arrive at the analyser position.

3.1 Scan Tailoring for ROTAX

The lower limit of the physical scan part mentioned above is defined by the used maximum incident neutron energy, whereas the upper limit is in general defined by technical limitations of the analyser crystal's motor drive, e.g. its maximum acceleration or power change.

The analyser rotation functions (eq. 4) are of course continuous and periodic in time and must be synchronised with the pulse repetition frequency of the neutron source. Thus, the physical scan time intervall will be followed by a region, in which the analyser crystal is being reset to its original angular position and corresponding speed and acceleration ready for the next pulse. We shall denote this particular time interval the repositioning scan part of a total motor revolution period.

The particular motion within this repositioning part is, however, not unique. Apart from technical limits for the motor's voltages and currents, the basic constraints are to match with the source's boundary conditions:

$$\int_0^{T=20\ ms} \dot{\omega}_A(t_A)\, dt_A = 0 \tag{5a}$$

$$\int_0^{T=20\ ms} \omega_A(t_A)\, dt_A = n\,2\pi \tag{5b}$$

We can make use of the symmetry of the analyser crystal with a mutliplicity m of its reflection plane, which allows the motor to turn only a fraction of $2\pi/m$ within the time period of one neutron frame. Thus, the average angular speed can be reduced by:

$$\frac{1}{T}\int_0^{T=20\ ms} \omega_A(t_A)\, dt_A = \frac{1}{m} \cdot 50\ Hz \tag{5c}$$

Fig. 2 illustrates the various choices of scan paths for physically turning the analyser crystal; the solid line denotes the fixed physical part, the dashed lines the quasi-arbitrary repositioning part.

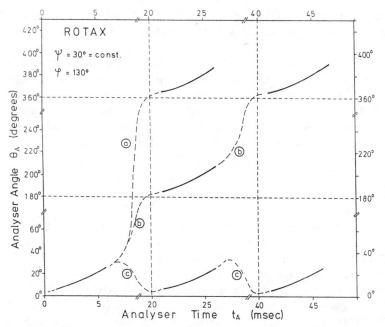

Fig. 2. Periodicity scheme of the analyser's rotation with various paths for repositioning assuming a) 2π-symmetry, b) π-symmetry and c) oscillating mode

For the physical part, however, the total resolution and intensity performce will require dynamic angular and time tolerances in positioning the analyser crystal restricted to approx. $\Delta\Theta_A \cong 0.1°$ and $\Delta t_A \cong 100\mu s$, respectively, for all types of scans (Tietze et al 1989), whereas in the repositioning part these severe requirements to the accuracy of the motor positioning can be released.

It turned out to be advantageous to construct the analyser rotation function with respect to a minimisation of the root mean square torque of the motor (Schmidt et al 1989). A polynomial representation of the order 7 for the motor's general angular function $\theta(t)$ gives a possible solution in the repositiong time interval $t_1 \le t \le t_2$:

$$\theta(t) = \sum_{n=0}^{7} a_n (t - t_1)^n \qquad (6)$$

where the coefficients can be determined from the boundary conditions:

$$\theta(t_2) = \Theta_A(t_2) ; \qquad \theta(t_1+T) = \Theta_A(t_1) + k\pi \qquad (7a)$$
$$\dot{\theta}(t_2) = \dot{\Theta}_A(t_2) ; \qquad \dot{\theta}(t_1+T) = \dot{\Theta}_A(t_1) \qquad (7b)$$
$$\ddot{\theta}(t_2) = \ddot{\Theta}_A(t_2) ; \qquad \ddot{\theta}(t_1+T) = \ddot{\Theta}_A(t_1) \qquad (7c)$$
$$\dddot{\theta}(t_2) = \dddot{\Theta}_A(t_2) ; \qquad \dddot{\theta}(t_1+T) = \dddot{\Theta}_A(t_1) \qquad (7d)$$

The integer value k (eq. 7a), which corresponds to the various choices of repositioning illustrated in fig. 2, is determined such that the torque minimises.

A desired scan with ROTAX, i.e. the desired rotational function of the motor drive with respect to a particular scan in the experimentally accessible (Q,ω)-space (physical part of the scan) and its corresponding requirements in the repositioning part of the scan, is then totally determined by the two analytical functions eq. (1) and eq. (6), or, more convenient for modern electronics, written as numerical arrays for $\Theta_A(t)$, $\dot{\Theta}_A(t)$ and $\ddot{\Theta}_A(t)$. These arrays are, in fact, the actual user-defined input values to the electronic control device described next.

3.2 Logic and Hardware Configuration

There is a spectrometer simulation program on the front-end computer (μVAX II) of the instrument, which the user will employ to design a particular scan and to test its feasibility. When a suitable and feasible scan is determined, the program creates arrays containing the desired input values for the analyser's angular position $\Theta_A(t)$, its angular speed $\dot{\Theta}_A(t)$ and acceleration $\ddot{\Theta}_A(t)$. They are digitized equi-distant in time and stored in a memory of the control electronics (cf. fig. 3). Throughout this article, we like to introduce two handy German keywords: "soll"-value for any kind of desired input quantity and "ist"-value for the corresponding actually measured quantity. Thus, the regulation logic is essentially a feed-back loop with a "ist"-versus-"soll" comparison.

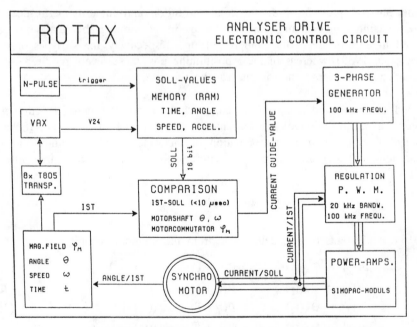

Fig. 3. Block-diagram of the analyser's drive electronic control.

When starting a run, the "soll"-arrays are transferred to the comparison unit, where they are translated into the appropriate electrical "soll"-data and trigger the triple-phase generator. There, the integral input signal is split into three phases for the electric currents and voltages, which are finally amplified by 12 MOSFET powermoduls SIMOPAC® BSM 191 F (2 for each polarity and phase) with integrated fast recovery epitaxial diodes (so-called FREDFETs®) to enable switching times of only 150 nsec at 1000 V and 28 A. It is neccessary to regulate the resulting high power electric current by means of a pulse width modulated power supply for the final amplifiers. A 100 kHz internal modulation frequency is used in order to be fast enough, whereas the bandwidth of 20 kHz of the pulse width modulator governs its effective power amplitude. The motor, which is a brushless, electronically commutated servo-motor from Polymotor Italiana S.p.A. Genova, Italy, converts the actually provided currents into an angular motion which is monitored by an internal resolver system. From this resolver we obtain a 14 bit wide signal for the actual angular position and 4 independent 90 degrees signals for absolute zero-point gauging. The time between two subsequent pulses from the resolver-to-digital converter is measured with an internal 10 MHz real-time clock to determine the angular speed and further derivatives. This provides then the comparison unit with the actual values ("ist"- values) of the motor shaft and the motor's rotational magnetic field, essential for the electronic commutation of the motor. Thus, the feed-back loop is closed.

The time-synchronisation with the neutron source is established in the "soll"-value memory by a pointer to the appropriate time-index of the "soll"-arrays, and it is activated with every neutron pulse trigger sent out from ISIS. The comparator thus knows the actual start time and resets its internal clock to zero. Further, two types of veto flags are set, if the "ist"-versus-"soll" comparison fails within certain software limits. One flag is set for a single event, to disable reading from the temporary buffers of the neutron detector data acquisition. The other one, the multi-event flag, is set if a series of single event flags occur and a reliable performance of the whole regulation and control circuit is no longer guaranteed. The operating program will pick this multi-event flag and suspend the operation and send a warning to the system and the user.

4. SCAN OPTIMISATION

The resolver "ist"-values are also read directly into a fast processing computer (not the μVAX!) without disrupting the operation and control of the motor (cf. fig. 3). This is to directly monitor the motor's performance and if neccessary manipulate the "soll"-input functions on the basis of the encoder output. This feedback method is based on linear response theory, where we regard the measured output- or "ist"-function A as a convolution of the unknown system's response function S with the well known input- or "soll"-function E:

$$A = S \otimes E \tag{8}$$

In general A is different from E and the question is how to modify $E \mapsto E'$ that

$$A' = S \otimes E' := E \tag{9}$$

Unfortunately we cannot assume, that S is a specific and uniform response function of the system. It rather depends on a number of undefined and environmental parameters like temperature, humidity, friction of the bearings, variations in the temporary power requirements and so on. In order to improve the scan performance a deconvolution of the system's response function has to be performed from time to time. The Fourier transformed representation of eq. (9) reads:

$$\tilde{A}' = \tilde{S}\,\tilde{E}' = (\tilde{A}/\tilde{E})\,\tilde{E}' = \tilde{E} \tag{10}$$

$\tilde{E}$ is well known, $\tilde{A}$ is to be measured, thus $\tilde{E}'$ is obtained from

$$\tilde{E}' = \tilde{E}\,\tilde{E}\,/\,\tilde{A} \tag{11}$$

and E' subsequently through a reverse Fourier transform of eq. (11), which is then the desired modified input function.

First tests of this scan optimisation method to the ROTAX analyser drive are very promising. As an example we illustrate (cf. fig 4a) the actual scan performance of a constant-energy scan ($\hbar\omega = 50$ meV) without any optimisation and refinement of the electronic regulation characteristics. The upper part of fig 4 shows the "ist"-minus-"soll" difference of the analyser's Bragg angle versus time taken for one complete turn of the motor. This time interval of 40 msec corresponds to two neutron frames of ISIS. Further the "soll"-curves of the angular speed and acceleration, together with their rather wobbly "ist"-curves, are presented in fig. 4. From fig. 4a it is obvious that the angular misfit of up to 17 degrees originates from the time delay of the system by approx. 2.5 msec. Fig. 4b presents the results of exactly the same scan after one optimisation cycle of deconvoluting the system function. It is obvious, that the angular accuracy and delay time shift have improved considerably. In principal, the method described above, can be repeated many times. In fact, it is advantageous to adjust first of all the zero'th order offset values $\Delta\Theta_0$ and Δt_0 in angle and time seperately, and to apply the Fourier transform afterwards. Fig. 5 illustrates schematically the algorithm being used throughout this whole procedure.

However, there are some disadvantages which should be taken into account:
a) The modified "soll"-scan is obtained by dividing the originally desired scan by an experimentally obtained array of "ist"-values, certainly carrying a number of statistical errors. They have to be minimised by statistical methods first.
b) Linear response theory does not at all account for non-linear effect, which certainly will occur in such a complex system, e.g. when the technical limits of the electric power control are approached. Non-linear effects, however, may be compensated by empirical convergence factors in the various feed-back cycles illustrated with fig. 5.
c) System transients or resonances, where the "ist"-array A (eq. 11) diverges, have to be avoided experimentally.

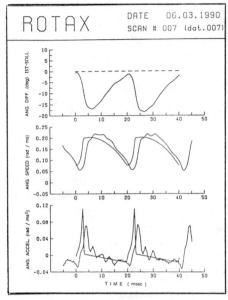
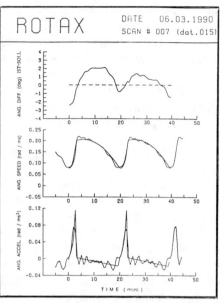

Fig. 4. Scan performance of a test scan with typical requirements:
a) angular difference "ist"-"soll", angular speed and acceleration versus time; without any refine-ment (see text).
b) angular difference "ist"-"soll", angular speed and acceleration versus time; after one loop of Fourier transform refinement (see text).

Nevertheless, the above presented method shows very promising results. We have not sufficient experience yet to answer the question, how often such a control optimisation and refinement loop actually has to be performed during an experimental scan? Is it sufficient to optimise once a day, or will we have to do the after every few motor revolutions?

If such an optimisation would be neccessary very frequently, it will impose quite severe requirements on the computing power and data transfer rates of the instrument computer: E.g. reading 7200 data points from the resolver at maximum speeds of

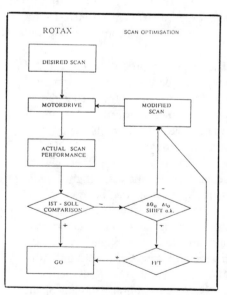

Fig. 5. Schematic diagram of the scan optimisation procedure.

100 rad/sec with 14 bit resolution for the angle and 9 bit for the velocity, we may end up with peak data rates of up to 2.2 Mbyte/sec to be thrown into a buffer's memory. Further, these data have then to be Fourier transformed twice within a few milli–seconds of time and subsequently fed back to the control electronics. Nowadays, fast computer components, e.g. accelerating systems to μVAX computers, like T805 transputers or amorphous array processors like the MAP4000 system, are commercially available. In principal, this should gives us the neccessary computing power.

5. FIRST RESULTS WITH NEUTRONS

With the technical configuration described in section 3.2 first neutron scattering experiments have been performed on a test assembly of ROTAX using the test beam line S9 of ISIS. A beam viewing the same moderator will also be used for the final installation of ROTAX. A Ni-powder sample has been used for this particular experiment. The Ni-powder sample was placed into the white incident neutron beam for selecting a number of scattered Bragg energies. Further, a cylindrically shaped Germanium single–crystal (1 cm $\emptyset$ and 3 cm hight) has been used to Bragg analyse the particular neutron energies. Two independent rotary detector arms have been installed, each equipped with 5 ^{3}He-counters covering 5 degrees of scattering angles. The Ge-crystal was mounted onto the motor shaft and we found that it was well balanced with respect to the motor axis. It was aligned with its $<01\bar{1}>$ crystallographic axis perpendicular to the scattering plane. Thus a variety of possible Ge-Bragg reflections could be used as analysing planes, e.g. (111), (022), (311), (400) etc. A schematical set-up of the test experiment is illustrated in fig. 6 showing the actual configuration for the first successful neutron scattering with ROTAX in the "lighthouse mode".

The aligment procedure of ROTAX may considered as a demonstration for the flexibility of the spectrome-ter: A powder spectrometer mode, where the detector is put in a straight-through position, i.e. $2\Theta_A = 0$, was used to determine precisely the actual spectrometers flight paths then, by putting the analyser crystal into the direct white beam, its scattering plane has been determined and oriented in a single crystal dif-fractometer mode. Finally, a PRIS-MA mode with the final geometry of fig. 6, but without rotation of the analyser crystal, has been used to determine the offset values $\Delta\Theta_0$ and

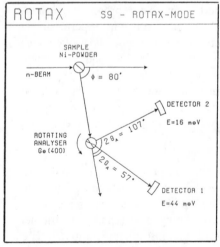

Fig. 6. The ROTAX test assem-bly on the ISIS S9 beamline

Δt_o, i.e. the $\Delta\Theta$-shift between the motor's and crystal's absolute angular zero and the time channel for which the scan optimisation procedure should be heading for. In fact, ROTAX incorporates all these different spectrometers within one.

Since we were restricted to two time-channels, because we only had two detector units, we scanned the two Ni-powder lines (200) and (311) in a simultaneous elastic scan, i.e. with zero energy transfer. Under the geometric conditions of this S9-test experiment with $\phi = 80°$ scattering angle and an incident moderator-to-sample flight path of 12.06 m and with sample-to-analyser-to-detector distances of 0.6 m and 0.585 m respectively, these Ni-powder lines (311) and (200) correspond to neutron energies of $E = 44$ meV and $E = 16$ meV at the correspon-ding time channels of 4.6 msec and 7.6 msec for the (modertor-to-detector) total time-of-flight. To observe these two energies within one ISIS-frame, the Ge-analyser crystal had to be positioned to the appropriate Bragg angles of $2\Theta_A = 57°$ and $2\Theta_A = 107°$ at time channels of $t_A = 4.5$ msec and $t_A = 7.5$ msec respectively (moderator-to-analyser neutron flight time), when using the Ge-(400) reflection. Fig. 7 shows the results obtained after ca. 20 minutes of counting time and fig. 8 illustrates the corresponding motor performance with the actual angular position and the corresponding derivatives, i.e. the angular speed and accelaration versus time. The dashed lines in fig. 8 indicate the two time channels, for which the performance of the system had been optimised (cf. section 4.).

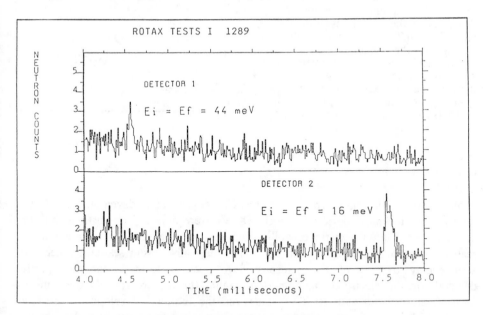

Fig. 7. Neutron scattering data obtained in the "lighthouse mode". The two different spectra show the intensities of the Ni-powder lines (311) and (200), see text.

The two peaks in fig. 7 correspond to the two neutron energies in the two detector units. Unfortunately, this signal has been lost throughout measuring during the night. The reason for that may seen in the weakness of the mechanical set-up of the test assembly, which was in fact not stiff enough to always withstand some very rough synchronised motor turns, when some out-of-phase frames of ISIS had reset immediately the instantaneous angular speed to its rather low starting value. A suitable error handling software will solve this problem for the next test period foreseen for summer 1990.

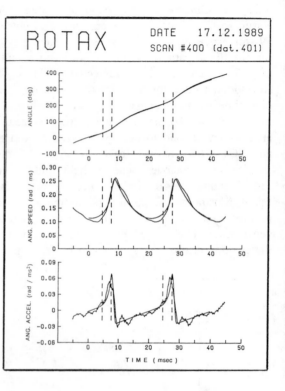

Fig. 8. Performance of the motor dynamics in the "lighthouse mode" (cf. figs. 6,7). "ist" and "soll" values of the angular position, speed and acceleration versus time with the dashed lines to indicate the time channels corresponding to the Ni-(311) and (200) Bragg reflections, see text.

6. CONCLUSION

After more than three years of research, design and development of the fundamental technical devices for ROTAX, we were able to demonstrate this first principal proof on the basic capability and performance of the instrument under realistic neutron scattering conditions. This time, only elastic scattering, i.e. with zero energy transfer of the neutrons, have been performed, but the specific demands on the electronics and motor dynamics (fig. 8) are roughly the same as for an inelastic scan with ROTAX. However, the technical development is by no means finished yet. Further improvements will enhance the accuracy and reliability of the ROTAX drive considerably. But it should be pointed out very clearly, that the recent success and performance is essentially based on the computer aided preparation and optimisation of the desired scan parameter, which govern and control the motor electronics for the analyser drive. At present we control the system with a bandwidth of 20 kHz of the amplifiers and this enables us to switch actually the full power of at present 2 kW within 50 μsec. And at least for the two crucial time channels, an angular accuracy of $\Delta 2\Theta_A$ better than 0.2° has been achieved under dynamically controled conditions. The success of the ROTAX spectrometer may be regarded as an example for a fruitful combination of modern electronics with modern computer techniques.

ACKNOWLEDGEMENT

We want to thank Mr. A J Chappell, Mr. A F Gilleard, Mr. E M Mott and Mr. S M Spurdle (RAL) for their help in the preparation period and their support during the test experiments on S9 at the Rutherford Appleton Laboratory. We also want to thank Dr. C Petrillo from CNR Frascati, Italy, for providing the excellent Ge-analyser crystal. This work has been performed under the terms of the "Rutherford-Würzburg" collaboration agreement and it has been funded by the German Federal Minister for Research and Technology (BMFT) under contract number 03-Ge2-Wue.

REFERENCES

Andreani C, Carlile C J, Cilloco F, Petrillo C, Sacchetti F, Stirling G C and
        Windsor C G 1987 Nucl. Inst and Meth. **A 254** 333
Geick R and Tietze H 1986 Nucl. Inst and Meth. **A 249** 325
Robinson R, Eckert J and Pynn 1985 Nucl. Inst. and Meth. **A 241** 312
Schmidt W, Tietze H and Geick R 1989 Physica **B 156 & 157** 554
Skilling J 1990 Proc. WONSDA II 1990 RAL, to be published
Steigenberger U, Hagen M, Caciuffo R and Sacchetti F 1990 RAL-90-004
        and 1990 Nucl. Inst. and Meth. to be published
Tajima K, Ishikawa Y, Kanai K, Windsor C G and Tomiyoshi S 1982 Nucl. Inst.
        and Meth **201** 491
Tietze H and Geick R 1987 Proc. ICANS IX 1986 CH-Villigen,
        SIN-rep 3-907998-01-4, 389
Tietze H, Schmidt W and Geick R 1989 Physica **B 156** 550

Inst. Phys. Conf. Ser. No 107: Chapter 6
Paper presented at Neutron Scatt. Data Anal. Conference, Rutherford Appleton, 1990

Analysis of quasielastic and inelastic scattering from crystal analyser instruments

A Smith, C J Carlile
ISIS pulsed neutron facility, Rutherford Appleton Lab,
Chilton, Didcot, Oxon OX11 0QX
M Prager
Institut für Festkörperforschung der Kernforschungsanlage
Jülich, Postfach 1913 D-5170 Jülich
and R M Richardson
Univ. of Bristol School of Chemistry, Cantock's Close,
Bristol BS8 1TS

ABSTRACT: The Wonderprogs, a set of quasielastic
scattering analysis programs, have been installed on the
RAL Neutron Division VAX. The programs have been embedded
within a menu system to make them more accessible to
inexperienced users, and modified for inverse geometry
spectrometers. A number of additional routines have been
written in order to aid users with the preparation of
control data.

A program for fitting inelastic data, ELF, is also
available. The data is fitted using a convolution of
either a number of Lorentzians or a number of Gaussians.

1. INTRODUCTION

Traditionally, quasielastic neutron spectroscopy has been
carried out on direct geometry time-of-flight (ToF)
spectrometers in which the neutron beam incident upon the
sample is monochromated using a chopper or crystal, and the
broadening of the monochromatic beam by dynamic processes
in the sample is determined by ToF over a known scattered
flight path to a large bank of detectors. The C1G
spectrometer at Herald and the 4H5 spectrometer at Dido
were the first examples of this tradition in Britain, which
culminated in the building of IN5, the four chopper
spectrometer, and IN6, the crystal-chopper time-focussing
instrument, both at ILL, Grenoble. It is fair to say that
these two spectrometers have dominated the field of
quasielastic spectroscopy for the last decade and a half.

The crystal analyser or inverse geometry spectrometer has
been a more recent development in comparison with the ToF
machines, but it has certain advantages. An inverse
geometry spectrometer requires a white incident beam, with
the final energy being selected by a crystal analyser
array, usually at or close to backscattering in order to
obtain the highest resolution. IN10, soon to be rebuilt as
IN10C, is the example most people are aware of. This
machine, with its resolution between 0.3 to 1μeV, succeeded
prototypes built at Garching and Jülich in Germany and is

conceptually the most complicated instrument in the field
of neutron scattering, with the possible exception of
neutron spin echo. Its white beam, with a maximum energy
transfer range of 15μeV either side of the elastic line, is
produced by Doppler shifting the incident neutrons whilst
they are Bragg reflected from a monochromator crystal
mounted on an oscillating piston. IN13, a later addition to
the crystal analyser stable at ILL, produces its white beam
by cycling the temperature of the monochromator in a
furnace. The obvious difficulty of generating a white beam
was partly the reason why crystal analyser instruments were
late in development and are still quite rare instruments,
particularly on reactor sources.

Pulsed sources have made it much easier to generate white
beams and crystal analyser instruments are naturally
adapted for such sources. The full advantages of crystal
analyser spectrometers, for example their wide energy
transfer range in neutron energy loss, can be exploited.
Energy levels in very cold samples can thus be
unequivocally determined.

In this paper we are concerned with the analysis of data
from the IRIS crystal analyser spectrometer at the ISIS
pulsed neutron facility. IRIS is designed for high
resolution quasielastic and inelastic spectroscopy and,
depending upon the analyser reflection chosen, resolutions
are available from 50μeV ((004) pyrolytic graphite) to 5μeV
((004) mica). With detectors at scattering angles from 15°
to 165° a Q-range of 0.08 Å^{-1} to 3.7 Å^{-1} can be sampled,
which is ideal for quasielastic studies.

The instrument is shown in Fig. 1. The sample sits at 36m
from the liquid hydrogen moderator on ISIS and is
illuminated by a white neutron beam whose time of arrival
at the sample is correlated with energy. A curved guide
maintains source intensity at the sample and a disc chopper
at 6.4m limits the wavelength band to avoid frame overlap.
A beam monitor just before the sample enables the incident
neutron spectrum to be measured. A typical incident
spectrum is shown in Fig. 2.

The graphite and mica analyser arrays are close to
backscattering geometry at 2θ = 175° to enhance resolution
and the analysed neutrons are detected in a 51-element zinc
sulphide scintillation detector covering scattering angles
from 15° to 165°. A 3-element diffraction detector is
situated at a scattering angle of 170° to enable a
simultaneous diffraction pattern of the sample to be
recorded. This proves particularly useful as a diagnostic
tool when studying the dynamics of samples near phase
transitions and for obtaining structural information on
unique samples prepared in-beam.

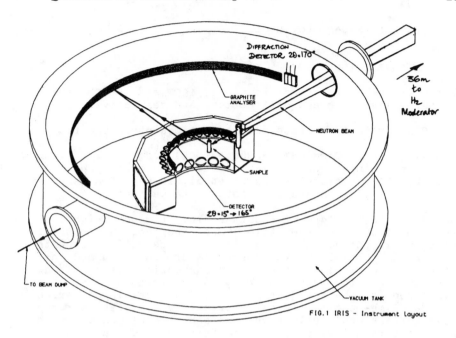

Fig. 1. A perspective view of the IRIS quasielastic scattering spectrometer showing the pyrolytic graphite analyser bank.

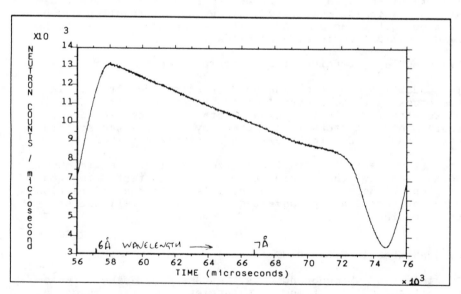

Fig. 2. The white spectrum incident upon the sample, expressed as a function of time of flight and wavelength. The range can be selected by changing the phase of the disc

chopper with respect to the ISIS pulse.

2. DATA CORRECTION

There are two correction routines, ICON and DEMON, which treat the inelastic and diffraction data respectively.

2.1 ICON – Data Correction for Inelastic Neutron Spectra

ICON is a Genie Command Language (GCL) program, run from within the Genie spectrum manipulation and display package used by ISIS instruments (David et al, 1986). ICON is used to correct IRIS raw data for instrumental effects.

The data, in the form of counts against time of flight, is read in from the raw file, and the incident monitor spectrum is corrected for variation of its efficiency with wavelength. The 51 analysed spectra are then summed into groups of a fixed number of neighbouring detectors as specified by the user. The grouped spectra are then normalised to the incident monitor. The data is converted from a time-of-flight scale to energy transfer for the selected analyser reflection, and can also be converted to the scattering law, $S(Q,\omega)$. In addition, ICON allows the correction of two runs in parallel so that one may be used as a background and subtracted from the other.

The resultant data is written to a number of workspaces within Genie for temporary storage, and if required can be written out to an external file in the user's area to be read back into Genie at a later date.

2.2 DEMON – Correction of Diffraction Spectra

In addition to the analysed neutron detectors, IRIS possesses a three-element diffraction detector, close to backscattering. Diffraction data is corrected for instrumental effects in the same way as analysed spectra, using DEMON, a GCL routine which is similar to ICON.

The raw data is corrected for monitor efficiency, normalised to the incident spectrum, corrected for the wavelength dependence of the efficiency of the diffraction detector, and finally converted from a time-of-flight scale to d-spacing.

3. QUASIELASTIC SCATTERING

Quasielastic broadening is caused by the random repositioning of atoms or molecules, by translation or reorientation, on an observable time scale determined by the instrumental resolution. In general the Heisenberg uncertainty principle indicates the range of time scales observable on a particular instrument.

$$\Delta E \Delta t \simeq \hbar$$

where ΔE is the instrumental energy resolution and Δt is the jump time optimally observed. On IRIS translational or reorientational diffusive jumps with characteristic times from 10^{-9} to 10^{-11} seconds are observable.

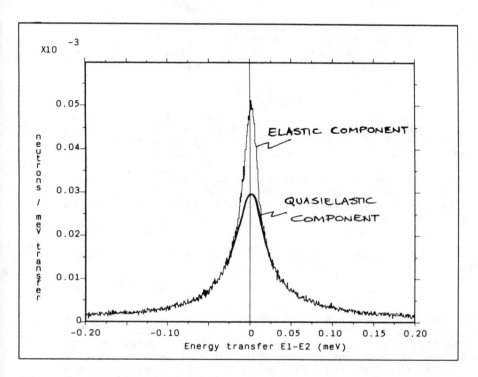

Fig. 3. A quasielastic spectrum from a polysaccharide gel taken at an intermediate Q as a function of energy transfer. The elastic and quasielastic components are indicated.

The quasielastic line will, in general, comprise two components, purely elastic and purely quasielastic, both lines being centred on an energy transfer $\hbar\omega = 0$. A typical spectrum is shown in Fig. 3. For purely translational diffusion the elastic component will be zero and the data analysis process will simply be used to determine energy width as a function of Q. In general, however, the aim of the data analysis process is to separate the elastic and the quasielastic components and to further subdivide the quasielastic scattering into its constituent parts, each of which depends upon the diffusional geometry. The scattering pattern observed $I(Q,\omega)$ is a convolution of the scattering law $S(Q,\omega)$ of the sample and the resolution $R(\omega)$ of the instrument.

$$I(Q,\omega) = S(Q,\omega) * R(\omega)$$

$S(Q,\omega)$ can be expressed, quite generally, by

$$S(Q,\omega) = A_0(Q)\delta(\omega) + \Sigma_n A_n(Q)L_n(\omega)$$

The first term, containing the delta function, is the purely elastic component and the succeeding terms are the quasielastic components. A structure factor, the EISF (Elastic Incoherent Structure Factor), can be defined as

$$EISF(Q) = \frac{A_0(Q)}{A_0(Q) + \Sigma A_n(Q)}$$

which contains geometrical information on the trajectories explored by the diffusing atoms over long times (ie long compared to the time window of the instrument).

It is the task of the data analysis process to determine the EISF and to obtain quantitative information on the widths and intensities of the various components within the quasielastic line.

4. THE WONDERPROGS - ANALYSIS OF QUASIELASTIC SCATTERING DATA

4.1 Introduction

The Wonderprogs are a package of quasielastic scattering analysis programs written by R.M.Richardson (1979), originally for use with data from ILL instruments. The programs have recently been modified for inverse geometry spectrometers and are installed on the RAL Neutron Division VAX, where they can currently be used in the analysis of IRIS data.

Two fitting programs, W8 and W8G, are supported at present. In each case a least squares fit with the Wonderprogs is made to a convolution of a model scattering function with the measured instrumental resolution. W8 makes an independent fit to each angle, whereas with W8G the fit is made to all angles simultaneously.

The model scattering function is defined by the user as a sum of terms whose shape, width, area and offset from zero may be individually specified. Parameters may be fixed or allowed to vary freely for the fit, and in addition pairs of parameters may be tied such that their values vary by a fixed ratio.

4.2 The fitting process

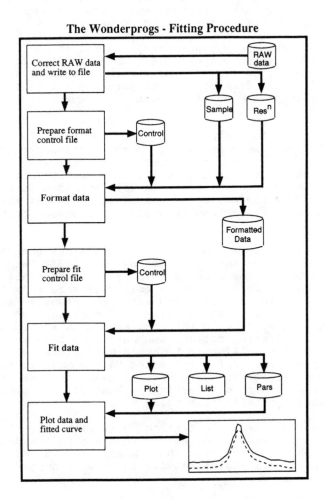

Fig. 4. A flow diagram of the Wonderprogs fitting process.

Fitting data with the Wonderprogs involves several stages of preparation, as illustrated in Fig. 4. In order to make the process more transparent to users a simple menu system has been written, and a number of additional routines have been installed to help with preparation of correctly formatted control files. The GCL routine ICON has also been installed within the menu system, in order to provide an interface between raw IRIS data and the fitting programs. The fitting process is described below.

Firstly, the required subset of the original sample and resolution data must be retrieved from the RAW file and, in the case of IRIS data, should be subsequently corrected using ICON. This is done in batch mode, in order to avoid the need to go into and out of Genie, and to make the process more transparent to inexperienced users.

The data must then be read into the first of the Wonderprogs, RTPW, which outputs the sample and resolution data in the format required by the fitting programs. This requires an additional file of control data and information about the experimental conditions such as temperature and detector angles.

Once the data is formatted, a second control file is required. This contains constraints on the fitting process and starting parameters for the model scattering law. The fitting programs require that this file is very precisely formatted and the meaning of the control data is unlikely to be clear to an inexperienced user, so it is at this point that the menu system and preparation programs prove most useful.

The program FITPREP allows the user to create a new control file, and prompts for the information required. Before writing to the file, the program displays a verification table through which the data can be edited if required. An existing control file can be read into the program and edited in the same way. The program's "intelligence" in leading the user through the data requirements in a step by step fashion, and offering options such as tying parameters only when appropriate, will help to avoid many pitfalls. In addition, a control file for a fit using W8 can quickly and easily be converted for use with W8G, allowing the user to find an approximate set of starting parameters using W8, and go on to make a detailed fit using W8G.

It should be noted that when fitting data with W8G, the total number of scattering angles or channels in the data may have to be reduced from that used with W8, as a simultaneous fit to all angles results in a large number of parameters.

The fitting programs are run in batch mode, as this proves to be quicker than running interactively. Times taken to complete a fit are of the order of a few minutes, although this is of course dependent upon the number of angles and channels in the data, as well as the number of iterations, and usage of the VAX cluster from moment to moment.

4.3 Output from the Wonderprogs

The fitting programs produce several output files. Firstly, an ASCII file of the sample data and fitted curve, which may be plotted at the terminal using either a GKS or UNIRAS graphics routine (both are provided). Typical UNIRAS output is shown in Fig. 5. Secondly, there are two listings of fitting parameters, one is a full description of the fitting process with initial and final values, plus other information; the other lists only the final values.

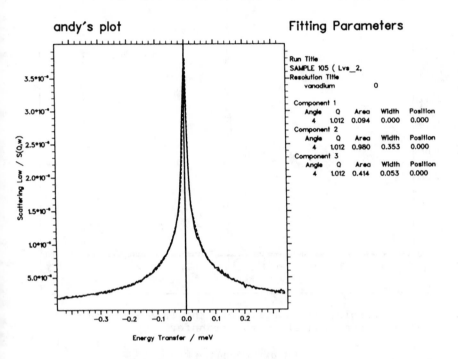

Fig. 5. Typical graphical display of data fitted using the Wonderprogs (UNIRAS graphics).

5. INELASTIC DATA ANALYSIS

ELF (Eleven Lorentzian Fit) is an inelastic fitting program which is specifically for the analysis of inelastic spectra with discrete lines such as quantum tunnelling or crystal field excitations. The program is derived from an original version written by R Osborn (1990). Up to 11 lines can be fitted simultaneously, convoluting the measured resolution function (see Fig. 6) with either Lorentzian or Gaussian inelastic lines.

The user has a wide choice of control in the program, ranging from allowing total freedom to being able to fix any of the parameters. Additionally, parameters can be tied so that, for example, the frequencies of tunnelling lines in energy gain and energy loss can be tied together but jointly allowed to vary. Intensities can also be tied.

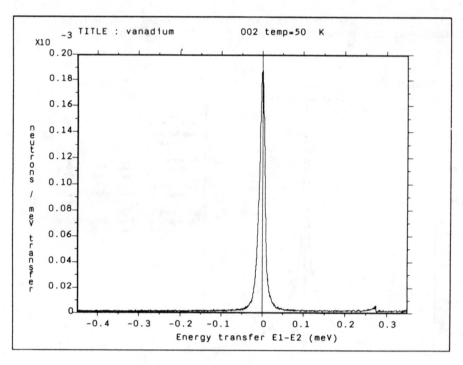

Fig. 6. The resolution function of the IRIS spectrometer, expressed as a function of energy transfer.

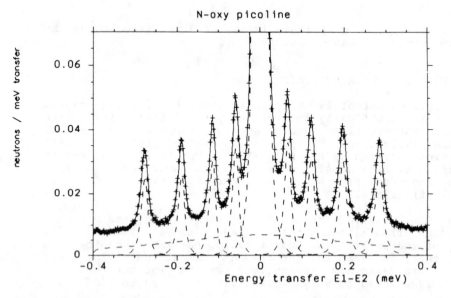

Fig. 7. The tunnelling spectrum of N-oxy picoline, fitted with ten lines using the ELF program (Carlile and Fillaux 1990)

A fit of the 8-line tunnelling spectrum of N-oxy picoline $C_5NH_4CH_3O$ is shown in figure 7. 2 further lines are included for the elastic line and to simulate the effect of thermal diffuse scattering in the graphite analyser.

6. CONCLUSION

The data correction and analysis programs described give a full set of routines for an inverse geometry white beam spectrometer on a pulsed source. The framework in which they are set allows extension to include routines presently not included such as multiple scattering corrections and more sophisticated graphical displays. It is hoped they will be found to be user friendly and extendable.

7. REFERENCES

W I F David ,M W Johnson, K J Knowles, C M Moreton-Smith, G D Crosbie, E P Campbell, S P Graham and J S Lyall, Rutherford Appleton Laboratory report 1986, RAL-86-102, PUNCH GENIE MANUAL Version 2.3 "A language for spectrum manipulation and display"

R M Richardson, Rutherford Appleton Laboratory report 1979, RL-79-095, "Notes on some quasielastic neutron scattering analysis programs on the Rutherford Laboratory IBM 360/195"

R Osborn, Rutherford Appleton Laboratory report (1990) in preparation

C J Carlile, F Fillaux (1990) private communication

Author Index

Anne M, *117*

Boudjada N, *117*

Carlile C J, *267*
Cilloco F, *185*
Clausen K N, *165*
Crennell K M, *83*

David W I F, *93*

Felici R, *185*
Figlarz M, *117*

Geick R, *253*
Ghosh R E, *233*
Gillon B, *101*

Hamilton W A, *223*
Hannon A C, *193*
Howe M A, *165*
Howells W S, *193*

Kearley G J, *245*
Keen D A, *165*

McGreevy R L, *165*
Moreton-Smith C, *69*

Pannetier J, *23*
Papoular R J, *101*
Penfold J, *213*
Piltz R O, *135*
Prager M, *267*
Pynn R, *45*

Rennie A R, *233*
Richardson R M, *267*
Rodriguez J, *117*

Samulowitz H, *253*
Schmidt W, *253*
Sibisi S, *1*
Silver R N, *45*
Sivia D S, *45, 223*
Skilling J, *1*
Smith A, *267*
Smith G S, *223*
Soper A K, *57, 193*
Steigenberger U, *253*

Tietze H, *253*
Tomkinson J, *245*

Wilson C C, *145*
Wolfers P, *127*

Subject Index

absorbtion, *135*
adsorption at interfaces, *214*
AgBr, *179*
amorphous samples, *193*
ATLAS, *193*
attenuation, *198*

Bayesian analysis, *45*
Bayes' theorem, *48, 105*
Boltzmann distribution, *165*
Boltzmann statistics, *24*
bond angle distribution, *175*

calibrant, *141, 233*
CCSL, *152*
CLIMAX, *245*
combination modes, *246*
configurational entropy, *172*
conjugated polymers, *223*
CONSTANT-Q, *253*
constrained model fitting, *219*
constraints, *127*
cost function, *25*
CRISP, *217*
crystal structure determination, *23, 34*
crystal structure from crystal-chemistry
    rules, *37*

D1A, D1B, D2B, *119*
D17, *233*
D20, *119*
data manipulation, *69*
density profile, *224*
detector deadtime, *198*
detector efficiency, *135*
differential cross-section, *193*
diffraction calibrants, *141*
diffuse scattering, *181*

EISF (elastic incoherent structure factor),
    272
energy landscape, *24*
error estimates, *1*
evidence, *1, 7*
extinction, *135*

false colour, *86*
figure-of-merit, *45*
FOCCOR, *138*
force constants, *245, 247*
Fourier, *53, 57, 94, 101, 102*

Fresnel's law, *214*
frustration, *24*

g(r), *168, 180, 195*
GENIE, *69, 83, 138, 270*
GKS, *83*
global optimization, *23*

hypothesis space, *49, 226*
HRPD, *93, 135*

ICON, *270*
image processing, *23*
inelastic neutron scattering, *245, 253, 268*
inelastic SANS components, *233*
inference, *1*
information content, *57*
instrumental design, *48*
instrumental resolution, *45*
interatomic potential, *12, 166, 172*
interfaces, *213, 223*
interfacial structures, *223*
interpolation theory, *4*
inverse geometry, *253, 267*
I(Q), *194*
IRIS, *268*
isotopic substitution, *168, 176, 213, 223*
iterative improvement, *23*

LAD, *193*
lattice-mode spectrum, *247*
least informative, *102*
least-squares, *94, 127, 245*
Lennard-Jones clusters, *36*
likelihood function, *48, 224*
lipid membranes, *223*
liquid bismuth, *182*
liquid diffraction, *185, 193*
liquid sulphur, *178*
local optimization, *23*

magnetic structures, *37, 103, 127*
magnetisation density, *101*
magnetism, *213*
Markov chain, *25, 165*
MAX, *253*
maximum entropy, *1, 49, 57, 93, 101, 107, 227*
mean-square displacements, *246*
Metropolis algorithm, *24, 165*
model fitting, *213*
modulated structures, *127*

molecular conformation, *36*
molecular dynamics, *165*
molecular modes, *245*
molecular structure determination, *23*
Monte Carlo (MC), *24, 58, 265*
morphology, *118*
multilayer optics, *213*
multiple scattering, *198*
MXD, *127*

neutron diffraction, *1, 93, 101, 117, 127, 135, 193*
neutron powder diffraction, *93, 117*
neutron reflectometry, *213, 223*
neutron refractive index, *213, 223*
non-crystalline neutron diffraction, *193*
non-linear least-squares, *213*
non-standard crystallographic refinement, *127*
normalization, *135, 185*
NUVU, *83*

optimal instrument design, *45*
orientation matrix, *149*
overtone modes, *246*

pair distribution function, *168, 180, 195*
partial radial distribution functions, *169*
partial structure factors, *168, 175*
Patterson entropy, *95*
Patterson function, *93, 94, 103*
PAW, *71*
peak searching, *150*
phaseless Fourier problem, *223*
phase transitions, *118*
phonon wings, *247*
Placzeck expansion, *187*
polarised neutron diffraction, *101*
position-sensitive detectors, *145*
posterior probability distribution function, *48, 224*
prior probability distribution, *1, 48, 224*
PRISMA, *253*
probabilistic hill climbing algorithm, *24*
probability theory, *1*
protein structures, *34*

quasielastic scattering, *267*
(Q, ω)-space, *253*

radial basis functions, *9*
radial distribution function, g(r), *168, 180, 195*
real-time neutron diffraction, *117*
reciprocal space, *145*
reciprocal space surveying, *149, 154*

reconstruction, *1*
refractive index profile, *213, 223*
Reverse Monte Carlo (RMC), *58, 165*
Rietveld technique, *93, 118*
ROTAX (ROTating Analyser crystal spectrometer), *253*

SANDALS, *193*
SANS, *233*
semi-conductor multilayers, *223*
SFLSQ, *152*
Shannon-Jaynes entropy, *107*
simulated annealing, *23, 58, 225*
single crystal diffraction, *145*
solid state transformation, *117*
spin density, *101*
S(Q), *185*
statistical cooling, *24*
STRAP, *118*
structural disorder, *165*
structural model, *165*
structure determination, *23, 34, 93, 127*
structure factors, *149, 166, 168, 185*
structure refinement, *119, 149*
surface chemistry, *213*
SXD, *145*
SXDRED, *152*
symmetry constraints, *127*

termination ripples, *197*
TFXA, *246*
thermaldiffractometry, *117*
topotactic dehydration, *117*
total cross-section, *196*
total scattering cross-section, *186*
transmission measurements, *233*
truncation effects, *101*

UB matrix, *149*
UNIRAS, *71, 83*

vibrational analysis, *245*
visualization, *69, 83, 251*

WONDERPROGS, *267*

x-ray reflectometry, *223*

Y_2O_3, *181*

$ZnCl_2$, *176*
ZrO_2, *181*

χ^2, *107*
$\delta(I)/I$ method, *151*